SpringerBriefs in Optimization

Series Editors

Panos M. Pardalos
János D. Pintér
Stephen M. Robinson
Tamás Terlaky
My T. Thai

SpringerBriefs in Optimization showcases algorithmic and theoretical techniques, case studies, and applications within the broad-based field of optimization. Manuscripts related to the ever-growing applications of optimization in applied mathematics, engineering, medicine, economics, and other applied sciences are encouraged.

For further volumes:
http://www.springer.com/series/8918

Remigijus Paulavičius • Julius Žilinskas

Simplicial Global Optimization

 Springer

Remigijus Paulavičius
Institute of Mathematics and Informatics
Vilnius University
Vilnius, Lithuania

Julius Žilinskas
Institute of Mathematics and Informatics
Vilnius University
Vilnius, Lithuania

ISSN 2190-8354   ISSN 2191-575X (electronic)
ISBN 978-1-4614-9092-0   ISBN 978-1-4614-9093-7 (eBook)
DOI 10.1007/978-1-4614-9093-7
Springer New York Heidelberg Dordrecht London

Library of Congress Control Number: 2013949407

Mathematics Subject Classification (2010): 90C26, 90C57, 52B11, 26B35, 90C90

Printed on acid-free paper

Springer is part of Springer Science+Business Media (www.springer.com)

Preface

Simplicial global optimization focuses on deterministic covering methods for global optimization partitioning the feasible region by simplices. Although rectangular partitioning is used most often in global optimization, simplicial covering has advantages shown in this book. The purpose of the book is to present global optimization methods based on simplicial partitioning in one volume. The book describes features of simplicial partitioning and demonstrates its advantages in global optimization.

A simplex is a polyhedron in a multidimensional space, which has the minimal number of vertices. Therefore simplicial partitions are preferable in global optimization when the values of the objective function at all vertices of partitions are used to evaluate subregions.

The feasible region defined by linear constraints may be covered by simplices and therefore simplicial optimization algorithms may cope with linear constraints in a delicate way by initial covering. This makes simplicial partitions very attractive for optimization problems with linear constraints.

There are optimization problems where the objective functions have symmetries which may be taken into account for reducing the search space significantly by setting linear inequality constraints. The resulted search region may be covered by simplices.

Applications benefiting from simplicial partitioning are examined in the book: nonlinear least squares regression, center-based clustering of data having one feature, and pile placement in grillage-type foundations. In the examples shown, the search region reduced taking into account symmetries of the objective functions is a simplex thus simplicial global optimization algorithms may use it as a starting partition.

The book provides exhaustive experimental investigation and shows the impact of various bounds, types of subdivision, and strategies of candidate selection on the performance of global optimization algorithms. Researchers and engineers will benefit from simplicial partitioning algorithms presented in the book: Lipschitz branch-and-bound, Lipschitz optimization without the Lipschitz constant. We hope the readers will be inspired to develop simplicial versions of other algorithms for global optimization and even use other non-rectangular partitions for special applications.

The book deals with theoretical, computational, and application aspects of simplicial global optimization. It is intended for scientists and researchers in optimization and may also serve as a useful research supplement for Ph.D. students in mathematics, computer science, and operations research.

The authors are very grateful to Prof. Panos Pardalos, Distinguished Professor at the University of Florida and Director of the Center for Applied Optimization, for his continuing encouragement and support. The authors highly appreciate Springer's initiative to publish SpringerBriefs on Optimization and the given opportunity to publish their book in this series. The authors would like to thank Springer's publishing editor Razia Amzad for guiding us to publication of the book.

Postdoctoral fellowship of R. Paulavičius is being funded by European Union Structural Funds project "Postdoctoral Fellowship Implementation in Lithuania" within the framework of the Measure for Enhancing Mobility of Scholars and Other Researchers and the Promotion of Student Research (VP1-3.1-ŠMM-01) of the Program of Human Resources Development Action Plan.

Vilnius, Lithuania   Remigijus Paulavičius
                                                                        Julius Žilinskas

Contents

Acronyms

n	Number of variables
$\mathbb{R}^n$	n-dimensional Euclidean space
$\mathbb{D}$	Feasible region
ε	Tolerance
x, y, z	Variables
$\mathbf{x}, \mathbf{y}, \mathbf{z}$	Vectors of variables
$f(\mathbf{x})$	Objective function
$\nabla f(\mathbf{x})$	Gradient of objective function $f(\mathbf{x})$
$F(\mathbf{x})$	Lower bounding function
f^*	Global optimum function value
$f(\mathbf{x}_{\text{opt}})$	ε-global optimum
$\mathbf{x}^*$	Global optimum vector
$\mathbf{x}_{\text{opt}}$	ε-global optimum vector
$\mathbb{S}$	Solution (subregion, optimum point)
$\mathbb{T}$	Finite set of points where the objective function value has been evaluated
$\mathbb{I}$	Subregion of feasible region
$\mathbb{L}$	Candidate set
$|\mathbb{L}|$	Cardinality of a candidate set
$\mathbf{v}$	Vertex of subregion
$\mathbb{V}(\mathbb{I})$	Set of vertices of subregion
$\mathbb{O}$	n-dimensional ball
LB	Lower bound for minimum
UB	Upper bound for minimum
R	Circumradius
D	Determinant
p	Number of processors
p, q	Norm index
s_p	Speedup
e_p	Efficiency

$\|\mathbf{x}\|_q$	q-norm, $(q \geq 1)$
$\|\mathbf{x} - \mathbf{y}\|_q$	Distance function
L_p	Lipschitz constant of objective function according to the p-norm
K	Lipschitz constant of derivatives
μ	Simple μ type Lipschitz bound
φ	Piyavskii type bound
ψ	Lipschitz bound based on the radius R of the circumscribed multidimensional sphere
$\mu_2^{1,2,\infty}$	μ_2 type Lipschitz bound with the $1, 2$, and ∞ norms
$\varphi^1 \psi^2 \mu_2^{2,\infty}$	Aggregate bound composed of φ, ψ, and μ_2 type bounds with different norms
$\overline{\varphi^1 \psi^2 \mu_2^{2,\infty}}$	Aggregate bound with vertex verification
r_{ψ^2/μ_2^2}	Ratio showing goodness of ψ^2 bound against μ_2^2 bound
$r(f^*)$	Search progress ratio
fe	Number of function evaluations
$t(s)$	Optimization time
TNS	Total number of simplices
MCL	Maximal size of candidate list

Chapter 1
Simplicial Partitions in Global Optimization

1.1 Covering Methods for Global Optimization

Many problems in engineering, physics, economics, and other fields may be formulated as optimization problems, where the optimal value of an objective function must be found [23, 55, 59, 110, 114, 134, 136]. The general global optimization problems solved by algorithms presented in this book can be written as follows:

$$\min \; f(\mathbf{x}), \quad f : \mathbb{R}^n \to \mathbb{R}$$
$$\text{s.t.} \; \mathbf{x} \in \mathbb{D} : \; g_1(\mathbf{x}) \le 0,$$
$$\vdots \tag{1.1}$$
$$g_m(\mathbf{x}) \le 0,$$
$$\mathbf{l} \le \mathbf{x} \le \mathbf{u},$$

where $\mathbb{D}$ is a nonempty feasible region, $g_1(\mathbf{x}), \ldots, g_m(\mathbf{x})$ are linear constraint functions, and $\mathbf{l} = (l_1, \ldots, l_n)$, $\mathbf{u} = (u_1, \ldots, u_n) \in \mathbb{R}^n$.

Most optimization problems considered in this book are constrained only by hyper-rectangular bounds on the variables. However, problems with linear inequality constraints will also be considered. For convergence reasons, we assume that the objective function is continuous in the neighborhood of the global minimizer. However, it can otherwise be nonlinear, non-differentiable, non-convex, and multimodal.

Besides the global optimum f^* one or all global optimizers $\mathbf{x}^* : f(\mathbf{x}^*) = f^*$ must be found or it must be shown that such a point does not exist. In this book we consider that $\mathbb{D}$ is compact and f is a Lipschitz continuous function, therefore the existence of $\mathbf{x}^*$ is assured by the well-known theorem of Weierstrass. Since maximization can be transformed into minimization by changing the sign of the objective function, we will consider only the minimization problems.

R. Paulavičius and J. Žilinskas, *Simplicial Global Optimization*,
SpringerBriefs in Optimization, DOI 10.1007/978-1-4614-9093-7_1,
© Remigijus Paulavičius, Julius Žilinskas 2014

Classification of global optimization methods was given in [136]:

- Methods with guaranteed accuracy:
  - Covering methods
- Direct methods:
  - Random search methods
  - Clustering methods
  - Generalized descent methods
- Indirect methods:
  - Methods approximating level sets
  - Methods approximating objective function

This book is focused on covering methods for global optimization. These methods partition the feasible region into subregions of a particular shape. The partitioning is stopped when the global minimizers are enclosed by small subregions achieving some prescribed accuracy.

Covering methods can detect and discard the subregions which do not contain the global minimum. A lower bound for the objective function over a subregion may be used to indicate the subregions which can be discarded. If guaranteed bounds are available, covering methods can ensure that a point $\mathbf{x}_{opt} \in \mathbb{D}$ is found such that $f(\mathbf{x}_{opt})$ differs from f^* by no more than a specified accuracy ε. Some covering methods are based on a lower bound constructed as a convex envelope of an objective function [33, 55, 77]. Lipschitz optimization is based on the assumption that the slope of an objective function is bounded [55, 59, 110, 134]. Interval methods estimate the range of an objective function over a subregion defined by a multidimensional interval using interval arithmetic [48, 92, 105].

Statistical models [146, 149] or heuristic estimates [83, 152] may also be used to evaluate subregions. Although guaranteed accuracy is lost in such a case, global optimization algorithms may be applied to solve "black box" optimization problems. In the "black box" situation, the values of an objective function are assumed to be given by an oracle, usually an objective function is given by means of a computer program and an analytical expression is not known, therefore the properties of the objective function are difficult to elicit.

A branch-and-bound technique can be used for managing the list of subregions and the process of discarding and partitioning. An iteration of a classical branch-and-bound algorithm processes a node in the search tree representing a not yet explored subregion of the feasible region. Each iteration has three main components: selection of a node to process, branching of the search tree by dividing the selected subregion, and pruning of the branches by discarding non-promising subregions. The rules of selection, branching, and bounding differ from algorithm to algorithm.

A general branch-and-bound algorithm for global optimization is shown in Algorithm 1. Before the cycle, the feasible region is covered by one or several partitions whose are added to the list of candidates $\mathbb{L}$.

Algorithm 1 General branch-and-bound algorithm

1: Cover $\mathbb{D}$: $\mathbb{L} \leftarrow \{\mathbb{L}_j | \mathbb{D} \subseteq \bigcup_{j=1}^{m} \mathbb{L}_j\}$ using **covering rule**
2: $\mathbb{S} \leftarrow \varnothing$, $UB(\mathbb{D}) \leftarrow \infty$
3: **while** $\mathbb{L} \neq \varnothing$ **do**
4:     Choose $\mathbb{I} \in \mathbb{L}$ using **selection rule**, $\mathbb{L} \leftarrow \mathbb{L} \setminus \{\mathbb{I}\}$
5:     **if** $LB(\mathbb{I}) < UB(\mathbb{D}) - \epsilon$ **then**
6:         Branch $\mathbb{I}$ into p subsets $\mathbb{I}_j$ using **branching rule**: $\mathbb{I} \subseteq \bigcup_{j=1}^{p} \mathbb{I}_j$
7:         **for all** $\mathbb{I}_j, j = 1, \ldots, p$ **do**
8:             Find $UB(\mathbb{I}_j \cap \mathbb{D})$ and $LB(\mathbb{I}_j)$ using **bounding rules**
9:             $UB(\mathbb{D}) \leftarrow \min(UB(\mathbb{D}), UB(\mathbb{I}_j \cap \mathbb{D}))$
10:             **if** $LB(\mathbb{I}_j) < UB(\mathbb{D}) - \epsilon$ **then**
11:                 **if** $\mathbb{I}_j$ may be a solution **then**
12:                     $\mathbb{S} \leftarrow \mathbb{I}_j$
13:                 **else**
14:                     $\mathbb{L} \leftarrow \mathbb{L} \cup \{\mathbb{I}_j\}$
15:                 **end if**
16:             **end if**
17:         **end for**
18:     **end if**
19: **end while**

There are three main and one additional selection strategies

- *Best first.* Select an element of $\mathbb{L}$ with the minimal lower bound. The candidate list must be prioritized structure, which can be implemented using a heap.
- *Depth first.* Select the youngest element of $\mathbb{L}$. A First-In-Last-Out structure is used for the candidate list which can be implemented using a stack. In some cases it is possible to implement this strategy without storing candidates as discussed in [147, 156, 158].
- *Breadth first.* Select the oldest element of $\mathbb{L}$. A First-In-First-Out structure is used for the candidate list which can be implemented using a queue.
- *Improved selection.* Based on heuristic [18, 68], probabilistic [25], or statistical [146, 149] criteria. In this strategy the candidate with the maximum criterion value is chosen [149].

The bounding rule describes how the bounds for the minimum of the objective function are found. The best currently found value of the objective function may be used as the upper bound for the minimum over the whole feasible region $UB(\mathbb{D})$. The lower bound for the minimum of the objective function over a considered subregion $LB(\mathbb{I})$ can be determined using convex envelopes, Lipschitz condition, or interval arithmetic.

The rules of covering and branching depend on the shape of the feasible region and the type of partitions used. Often feasible regions of global optimization problems are hyper-rectangles. Partitions may be hyper-rectangular, simplicial, hyper-conic, or hyper-spherical. All interval and most of Lipschitz global optimization branch-and-bound algorithms use hyper-rectangular partitions. Example rules of covering rectangular feasible region and branching are shown in Fig. 1.1: rectangular partitions are shown in the first row, simplicial in the second and

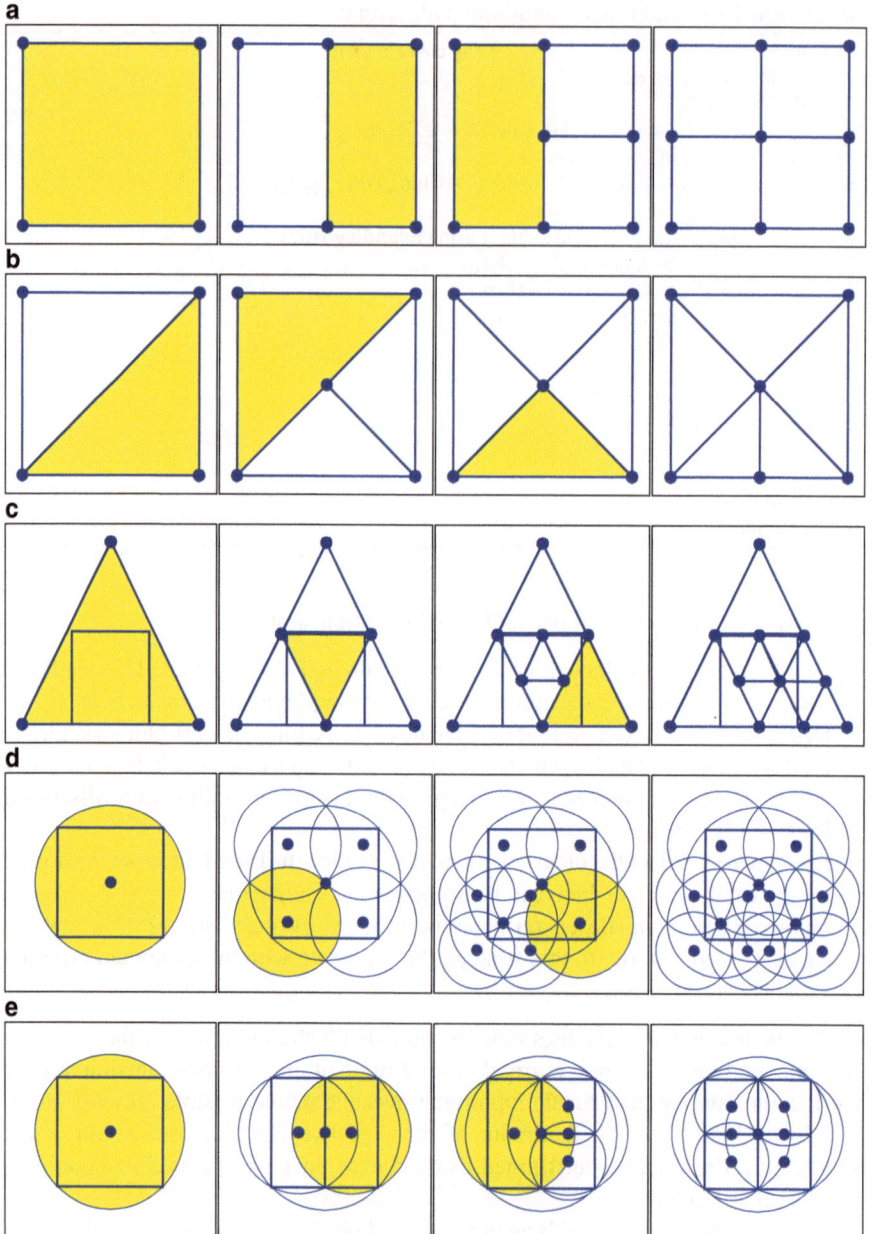

Fig. 1.1 Example rules of covering rectangular feasible region and branching: (**a**) rectangles, (**b**) irregular simplices, (**c**) regular simplices, (**d**) disks, (**e**) disks/rectangles

third rows, and spherical in the fourth and fifth rows. The first column shows covering rules and the others show branching. Partitions obtained by branch-and-bound algorithms for global optimization differ from those used in combinatorial optimization in that the number of possible partitions is infinite and that partitions overlap.

Initial covering is simple when the feasible region and partitions are of the same type : $\mathbb{L} = \{\mathbb{D}\}$ (see Fig. 1.1a). Covering of a hyper-rectangular feasible region by hyper-spheres causes over-covering as well as overlapping of spheres themselves (see Fig. 1.1d, e). The use of regular simplices causes over-covering of a hyper-rectangular feasible region and non-overlapping branching is not known in more than two dimensions (see Fig. 1.1c). The use of irregular simplices enables non-over-covering of feasible region as well as branching with non-overlapping interior (see Fig. 1.1b).

Although hyper-rectangular partitions are usually applied in global optimization, other types of partitions may be more suitable for some problems. Advantages and disadvantages of simplicial partitions are shown in this chapter.

1.2 Simplicial Partitioning

An n-simplex $\mathbb{I} = [\mathbf{v}_1, \ldots, \mathbf{v}_{n+1}]$ is the convex hull of a set of $(n+1)$ affinely independent points (vectors) $\mathbf{v}_1, \ldots, \mathbf{v}_{n+1} \in \mathbb{R}^n$, i.e.

$$[\mathbf{v}_1, \ldots, \mathbf{v}_{n+1}] = \left\{ \sum_{i=1}^{n+1} \theta_i \mathbf{v}_i \mid \theta_i \geq 0, i = 1, \ldots, n+1, \sum_{i=1}^{n+1} \theta_i = 1 \right\}.$$

A one-simplex is a segment of line, a two-simplex is a triangle, and a three-simplex is a tetrahedron (see Fig. 1.2). A simplex is a polyhedron in n-dimensional space, which has the minimal number of vertices $(n+1)$. Therefore simplicial partitions are preferable in global optimization when the values of an objective function at all vertices of partitions are used to evaluate subregions.

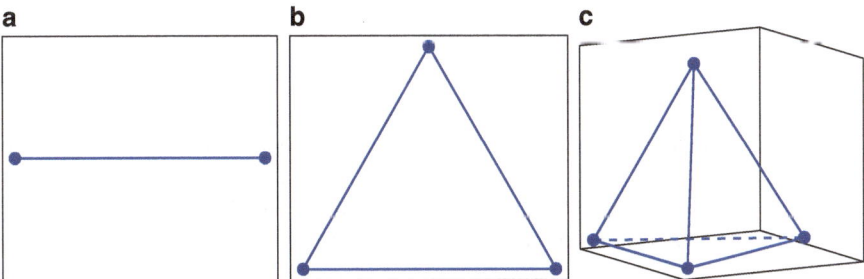

Fig. 1.2 n-simplices: (**a**) segment of line ($n = 1$), (**b**) triangle ($n = 2$), (**c**) tetrahedron ($n = 3$)

We define the sets of vertices ($\mathbb{V}(\mathbb{I})$) and edges ($\mathbb{E}(\mathbb{I})$) of a simplex $\mathbb{I}$ as follows:

$$\mathbb{V}(\mathbb{I}) = \{\mathbf{v}_1, \ldots, \mathbf{v}_{n+1}\}, \quad \mathbb{E}(\mathbb{I}) = \left\{ [\mathbf{v}_i, \mathbf{v}_j] \mid i, j = 1, \ldots, n+1, i < j \right\}.$$

The diameter ($\delta(\mathbb{I})$) of a simplex $\mathbb{I}$ is defined as

$$\delta(\mathbb{I}) = \max\{\|\mathbf{x} - \mathbf{y}\| \mid \mathbf{x}, \mathbf{y} \in \mathbb{I}\} = \max\{\|\mathbf{v}_i - \mathbf{v}_j\| \mid [\mathbf{v}_i, \mathbf{v}_j] \in \mathbb{E}(\mathbb{I})\}\}.$$

Definition 1.1. Let $\mathbb{I} \subset \mathbb{R}^n$ be a simplex and let $\mathbb{L} = \{\mathbb{I}_i\}$ be a finite set of nonempty simplices such that

$$\mathbb{I} = \bigcup_{\mathbb{I}_i \in \mathbb{L}} \mathbb{I}_i, \text{ and } \mathbb{I}_i \cap \mathbb{I}_j = \text{bnd } \mathbb{I}_i \cap \text{bnd } \mathbb{I}_j \quad \text{for all } \mathbb{I}_i, \mathbb{I}_j \in \mathbb{L}, \mathbb{I}_i \neq \mathbb{I}_j$$

where "bnd" denotes the boundary of a simplex, then we say that $\mathbb{L}$ is a simplicial partition of $\mathbb{I}$.

The convergence condition of branch-and-bound methods depends on the bounds and on the behavior of nested sequences of partition sets. We say that a sequence of simplices $\{\mathbb{I}_i \mid i \in \mathbb{N}\}$ is nested if $\mathbb{I}_{i+1} \subseteq \mathbb{I}_i$ for all i. Then for all i we have that $\delta(\mathbb{I}_{i+1}) \leq \delta(\mathbb{I}_i)$.

Definition 1.2. An infinite sequence of n-simplices $\mathbb{I}_i \in \mathbb{R}^n$ satisfying $\mathbb{I}_{i+1} \subseteq \mathbb{I}_i$ is called exhaustive if there is a point s such that

$$\lim_{i \to \infty} \mathbb{I}_i = \bigcap_{i=1}^{\infty} \mathbb{I}_i = \{s\}. \tag{1.2}$$

Exhaustiveness of $\{\mathbb{I}_i\}$ is equivalent to $\lim_{i \to \infty} \delta(\mathbb{I}_i) = 0$. Subdivision of simplices during a search for the global minimum is carried out by means of partitioning a simplex into a finite number of sub-simplices.

Definition 1.3. Let $\mathbb{I}$ be an n-simplex with vertex set $\mathbb{V}(\mathbb{I}) = \{\mathbf{v}_1, \ldots, \mathbf{v}_{n+1}\}$. Choose a point $\mathbf{w} \in \mathbb{I}, \mathbf{w} \notin \mathbb{V}(\mathbb{I})$ which is uniquely represented by

$$\mathbf{w} = \sum_{i=1}^{n+1} \lambda_i \mathbf{v}_i, \quad \lambda_i \geq 0 \ (i = 1, \ldots, n+1), \quad \sum_{i=1}^{n+1} \lambda_i = 1 \tag{1.3}$$

and for each i such that $\lambda_i > 0$ form the simplex $\mathbb{I}(i, \mathbf{w})$ obtained from $\mathbb{I}$ by replacing the vertex $\mathbf{v}_i$ by $\mathbf{w}$, i.e., $\mathbb{I}(i, \mathbf{w}) = [\mathbf{v}_1, \ldots, \mathbf{v}_{i-1}, \mathbf{w}, \mathbf{v}_{i+1}, \ldots, \mathbf{v}_{n+1}]$. This subdivision is called a radial subdivision, also referred to in the literature as ω-subdivision, which provides another commonly used method for partitioning [8, 52, 137, 138].

The number of sub-simplices $\mathbb{I}(i, \mathbf{w})$ is equal to $d + 1$ when the smallest face containing $\mathbf{w}$ has dimension d. When $\mathbf{w}$ lies on an edge of $\mathbb{I}$, we obtain two sub-

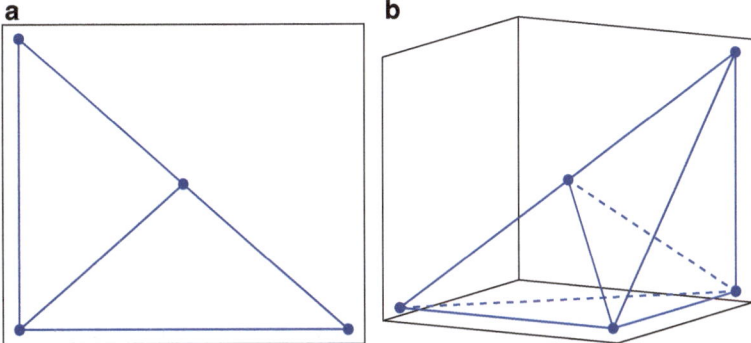

Fig. 1.3 Bisecting a simplex through the middle of the longest edge: (**a**) $n = 2$, (**b**) $n = 3$

simplices. Such a subdivision strategy is often called a bisection of $\mathbb{I}$ and was commonly used in the literature [10,54,65]. However it was showed in [22] that such a subdivision strategy does not have much freedom in the choice of the edge of a simplex $\mathbb{I}$ if we wish to guarantee the exhaustivity of the sequence of partitions. This is because, with this method, the same edge in two different sub-simplices should really be considered as two separate edges due to the fact that they are bisected separately. For this reason some authors simultaneously partition all sub-simplices with a common edge [22,44] which gives more freedom in terms of exhaustivity.

Proposition 1.1. *A sequence of simplices* $\mathbb{I}_i = [\mathbf{v}_{i_1}, \ldots, \mathbf{v}_{i_{n+1}}], i = 1, 2, \ldots$ *with the longest edges* $[\mathbf{v}_{i_1}, \mathbf{v}_{i_2}]$ *is exhaustive when* $\mathbb{I}_{i+1}$ *is constructed from* $\mathbb{I}_i$ *by bisection using points* $\mathbf{w}_i \in [\mathbf{v}_{i_1}, \mathbf{v}_{i_2}]$ *satisfying*

$$\mathbf{w}_i = \lambda_i \mathbf{v}_{i_1} + (1 - \lambda_i)\mathbf{v}_{i_2}, \quad 0 < c \leq \lambda_i \leq \frac{1}{2} \qquad (1.4)$$

for some fixed number $0 < c < \frac{1}{2}$.

The proof of this proposition can be found in [55]. When $\mathbf{w}_i$ is the midpoint of a longest edge of $\mathbb{I}$ ($\lambda_i = \frac{1}{2}$), we ensure that the longest edge of the sub-simplices is not more than two times longer than other edges. The examples of such a subdivision are shown in Fig. 1.3.

A triangle (two-dimensional simplex) may be divided into four similar triangles, a right equilateral triangle may be divided into two similar triangles, see Fig. 1.4. When a regular triangle is subdivided into four all of them are regular too. Such a subdivision ensures all triangles are regular. Unfortunately such a strategy of subdivision is not known in $n > 2$. Experiments in [151] have shown that subdivision by a hyper-plane passing through the middle point of the longest edge and the vertices not belonging to the longest edge is preferable over division into several simplices.

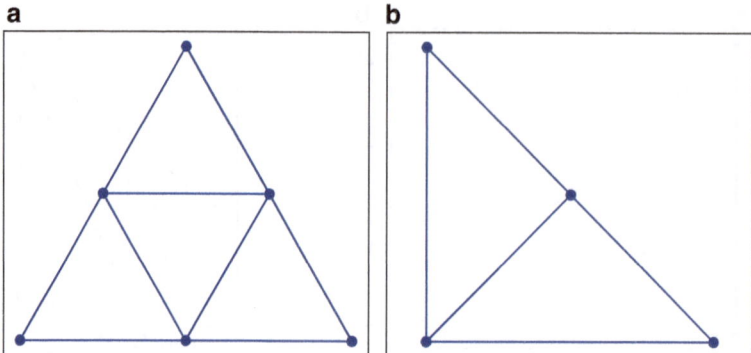

Fig. 1.4 Subdivision of triangles into similar triangles: (**a**) into four, (**b**) right equilateral into 2 right equilateral

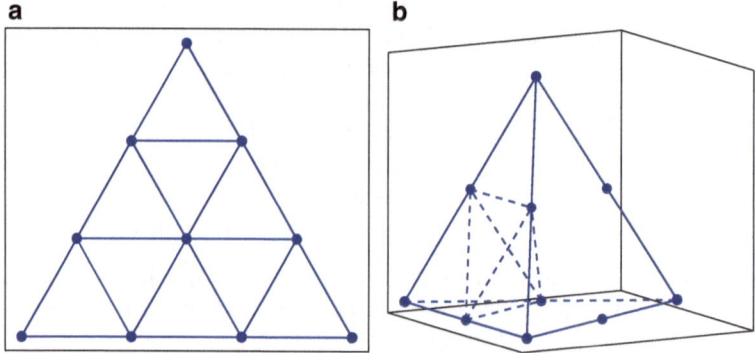

Fig. 1.5 Subdivision by means of an edgewise subdivision: (**a**) into 3^2, (**b**) into 2^3

Two algorithms for the edgewise subdivision of an n-simplex into k^n simplices (with the base k) of the same volume and shape characteristics was proposed in [26, 43]. An abacus model of a simplex, called the color scheme, is proposed in [26] to subdivide an n-simplex into k^n n-simplices. In [43], the authors present a ready-to-implement algorithm for automatic generation of color schemes for a general k^n case. In the case of $k^n = 2^2$ edgewise subdivision provides the same results as in Fig. 1.4a. The edgewise subdivisions of a triangle into $k^n = 3^2$ is shown in Fig. 1.5a and of a tetrahedron into $k^n = 2^3$ is shown in Fig. 1.5b.

Exhaustiveness guarantees convergence. More sophisticated exhaustive subdivision procedures were proposed in [56, 57]. In a recent article [22] the conditions were investigated when we can guarantee that the sequence of simplicial partitions is exhaustive.

Proposition 1.1 suggested the bisection of the longest edge without taking into account the behavior of the objective function. Experiments in [56] showed that other subdivision rules might lead to better results, although exhaustiveness cannot be ensured.

1.3 Covering a Hyper-Rectangle by Simplices

Application of simplicial partitions on a hyper-rectangular feasible region requires covering by simplices. This is a challenge. There are two main covering strategies of a hyper-rectangle by simplices: over-covering and face-to-face vertex triangulation.

Using the first strategy, a hyper-rectangle is covered by one simplex. One version is to fit a hyper-rectangle into a simplex matching a vertex. One vertex of the hyper-rectangle and one vertex of the simplex are matched, edges of the simplex from this vertex include edges of the hyper-rectangle from this vertex, and the opposite vertex of the hyper-rectangle is placed on the opposite face of the simplex. Two-and three-dimensional examples are shown in Fig. 1.6. Such a covering is not unique. Implementation usually aims at minimization of the over-covering.

Consider a simplex $\mathbb{I} = [(0, 0, \ldots, 0), (1, 0, \ldots, 0), (0, 1, \ldots, 0), \ldots, (0, 0, \ldots, 1)]$ in n-dimensional space is given. Let's find the hyper-rectangle of the largest hyper-volume which can be covered by this simplex by matching a vertex. One vertex of the hyper-rectangle is at the origin $(0, 0, \ldots, 0)$ and the opposite one $(x_1, x_2, \ldots, x_n)$ should be on the standard simplex $[(1, 0, \ldots, 0), (0, 1, \ldots, 0), \ldots, (0, 0, \ldots, 1)]$ and, therefore, should satisfy

$$\sum_{i=1}^{n} x_i = 1.$$

Let us formulate an optimization problem to find the maximal hyper-volume of such a hyper-rectangle. The variables are the coordinate values of the vertex opposite to the vertex at the origin and together sizes of the hyper-rectangle. Let us maximize the hyper-volume with the constraint that the vertex should be on the standard simplex:

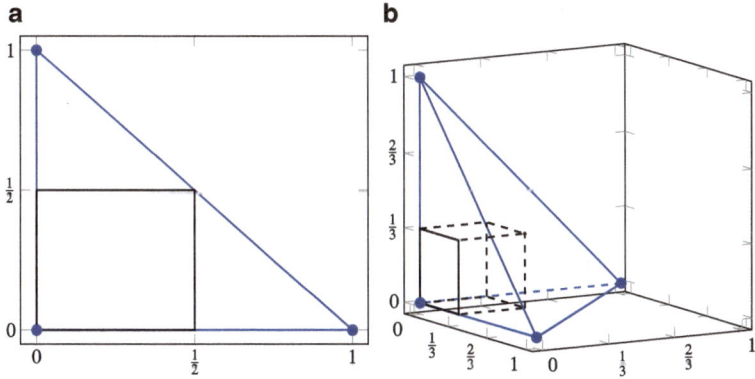

Fig. 1.6 Over-covering of a hyper-rectangle by matching a vertex: (**a**) $n = 2$, (**b**) $n = 3$

$$\max \ \{V(\mathbf{x}) = \prod_{i=1}^{n} x_i\}$$

$$\text{s.t.} \ \sum_{i=1}^{n} x_i = 1,$$

$$x_i \geq 0.$$

We can reduce one variable, say x_n, by expressing it from the equality constraint:

$$x_n = 1 - \sum_{i=1}^{n-1} x_i.$$

Substituting the expression to $V(\mathbf{x})$ we result in a problem with an objective function

$$V(\mathbf{x}) = \prod_{i=1}^{n-1} x_i \left(1 - \sum_{i=1}^{n-1} x_i \right).$$

Let us find stationary points of $V(\mathbf{x})$. Partial derivatives

$$\frac{\partial V}{\partial x_i} = \prod_{j=1, j \neq i}^{n-1} x_j \left(1 - 2x_i - \sum_{j=1, j \neq i}^{n-1} x_j \right)$$

must be zero at stationary points. We are not interested in points where $x_i = 0$ since then the hyper-volume is zero. Therefore, we are interested in a point where

$$2x_i + \sum_{j=1, j \neq i}^{n-1} x_j = 1, \ \forall i.$$

Such a point can be found solving a system of linear equations

$$(\mathbf{I} + \mathbf{E})\mathbf{x} = \mathbf{e},$$

where $\mathbf{I}$ is an identity matrix and $\mathbf{E}$ is a matrix of ones, both of the size $(n-1) \times (n-1)$; $\mathbf{x}$ is a vector of $n-1$ unknowns and $\mathbf{e}$ is a vector of ones of the same size. The solution of the system of equations is

$$\mathbf{x} = (\mathbf{I} + \mathbf{E})^{-1}\mathbf{e} = (\mathbf{I} - \frac{1}{n}\mathbf{E})\mathbf{e} = \mathbf{e} - \frac{n-1}{n}\mathbf{e} = \frac{1}{n}\mathbf{e}.$$

From the expression of x_n we can get that it is also of the same value, what means all variables at the solution are equal to

$$x_i^* = \frac{1}{n}, \ \forall i$$

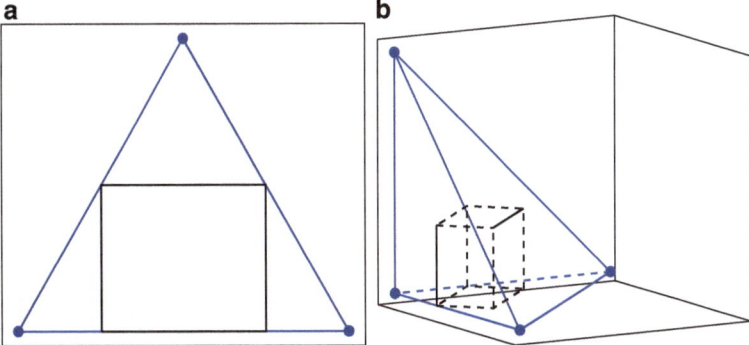

Fig. 1.7 Over-covering of a hyper-rectangle by placing its face on a face of a simplex: (**a**) $n = 2$, (**b**) $n = 3$

and the maximum volume is

$$V^* = n^{-n}.$$

Therefore, the ratio of volumes of this simplex and the largest hyper-rectangle covered by it is

$$\frac{V_s}{V^*} = \frac{n^n}{n!} \approx \frac{n^n}{\sqrt{2\pi n}\left(\frac{n}{e}\right)^n} = \frac{e^n}{\sqrt{2\pi n}}.$$

This means that the ratio grows exponentially with n.

Another version of over-covering is to fit a hyper-rectangle into a simplex placing a face of the hyper-rectangle on a face of the simplex. Examples of such coverings are shown in Fig. 1.7. In such a case, covering using a regular simplex is possible.

The second covering strategy of a hyper-rectangle is face-to-face vertex triangulation: it is subdivided into n-simplices, whose vertices are also the vertices of the feasible region, see Fig. 1.8. A rectangle may be vertex triangulated into two triangles. A three-dimensional rectangle may be vertex triangulated into five simplices as shown in Fig. 1.8b. However such a triangulation is not general—it is not known in the case $n > 3$.

There are general (any dimensional) ways for face-to-face vertex triangulation. We call one of the ways by combinatorial triangulation [135]. The algorithm for combinatorial triangulation of a hyper-rectangle is shown in Algorithm 2. Here d_{j1} and d_{j2} represent the ends of interval of jth variable defining hyper-rectangular feasible region; v_{ij} represents jth coordinate of ith vertex of the current simplex. The approach is deterministic and based on enumeration of permutations of $\{1, \ldots, n\}$. The number of simplices is known in advance and equal to $n!$. All simplices are of equal hyper-volume, i.e., $1/n!$ of the hyper-volume of the

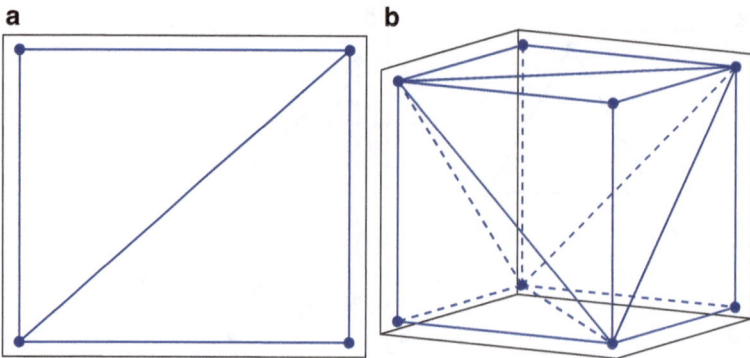

Fig. 1.8 Examples of face-to-face vertex triangulation by the smallest number of simplices: (a) $n = 2$, (b) $n = 3$

Algorithm 2 Combinatorial triangulation of hyper-rectangle

1: **for** τ = one of all permutations of $\{1, \ldots, n\}$ **do**
2:    **for** $j = 1, \ldots, n$ **do**
3:       $v_{1j} \leftarrow d_{j1}$
4:    **end for**
5:    **for** $i = 1, \ldots, n$ **do**
6:       **for** $j = 1, \ldots, n$ **do**
7:          $v_{(i+1)j} \leftarrow v_{ij}$
8:       **end for**
9:       $v_{(i+1)\tau_i} \leftarrow d_{\tau_i 2}$
10:   **end for**
11: **end for**

---

hyper-rectangle. The diagonal of the hyper-rectangle is an edge of all simplices. By adding just one point at the middle of the diagonal of the hyper-rectangle each simplex may be additionally subdivided into two.

For example, a unit cube is combinatorially triangulated into six simplices:

$$\tau = (1, 2, 3), \ \mathbb{I}_\tau = [(0, 0, 0), (1, 0, 0), (1, 1, 0), (1, 1, 1)],$$

$$\tau = (1, 3, 2), \ \mathbb{I}_\tau = [(0, 0, 0), (1, 0, 0), (1, 0, 1), (1, 1, 1)],$$

$$\tau = (2, 1, 3), \ \mathbb{I}_\tau = [(0, 0, 0), (0, 1, 0), (1, 1, 0), (1, 1, 1)],$$

$$\tau = (2, 3, 1), \ \mathbb{I}_\tau = [(0, 0, 0), (0, 1, 0), (0, 1, 1), (1, 1, 1)],$$

$$\tau = (3, 1, 2), \ \mathbb{I}_\tau = [(0, 0, 0), (0, 0, 1), (1, 0, 1), (1, 1, 1)],$$

$$\tau = (3, 2, 1), \ \mathbb{I}_\tau = [(0, 0, 0), (0, 0, 1), (0, 1, 1), (1, 1, 1)],$$

where τ represents a permutation and the corresponding simplex $\mathbb{I}$ is represented as the convex hull of vertices. Examples of combinatorial triangulation of two- and three-dimensional hyper-rectangles are shown in Fig. 1.9.

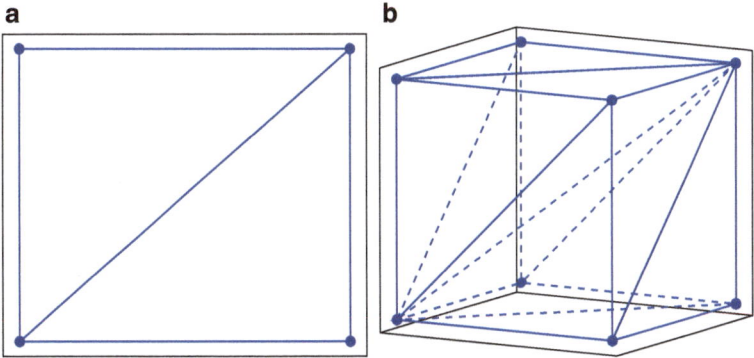

Fig. 1.9 Examples of combinatorial triangulation of hyper-rectangles: (**a**) $n = 2$, (**b**) $n = 3$

Algorithm 3 Parallel algorithm for combinatorial triangulation

---

1: **for** $k = \lfloor n!rank/size \rfloor$ **to** $\lfloor n!(rank + 1)/size \rfloor - 1$ **do**
2:    **for** $j = 1, \ldots, n$ **do**
3:        $\tau_j \leftarrow j$
4:    **end for**
5:    $c \leftarrow 1$
6:    **for** $j = 2, \ldots, n$ **do**
7:        $c \leftarrow c(j - 1)$
8:        swap $\tau_{j - \lfloor k/c \rfloor \% j}$ with τ_j
9:    **end for**
10:   **for** $j = 1, \ldots, n$ **do**
11:       $v_{1j} \leftarrow d_{j1}$
12:   **end for**
13:   **for** $i = 1, \ldots, n$ **do**
14:       **for** $j = 1, \ldots, n$ **do**
15:           $v_{(i+1)j} \leftarrow v_{ij}$
16:       **end for**
17:       $v_{(i+1)\tau_i} \leftarrow d_{\tau_i 2}$
18:   **end for**
19: **end for**

As the number of simplices for combinatorial triangulation is known in advance, efficient parallel enumeration of all simplices may be performed. Factoradic numbers can be used to assign unique numbers to permutations, such that given a factoradic of k one can quickly find the corresponding permutation. A parallel algorithm for combinatorial triangulation based on factoradic numbers is given in Algorithm 3, where $size$ is the number of processors, $rank$ is the number of the current processor from 0 to $size - 1$, and τ is the permutation of $\{1, \ldots, n\}$ corresponding to the current simplex. Such an algorithm may be considered for covering of hyper-rectangular feasible regions and initial distribution of work in parallel branch-and-bound algorithms with simplicial partitions.

Another triangulation of the feasible region with irregular simplices is proposed in [44]. This triangulation (see Fig. 1.10) is based on the vertices of the

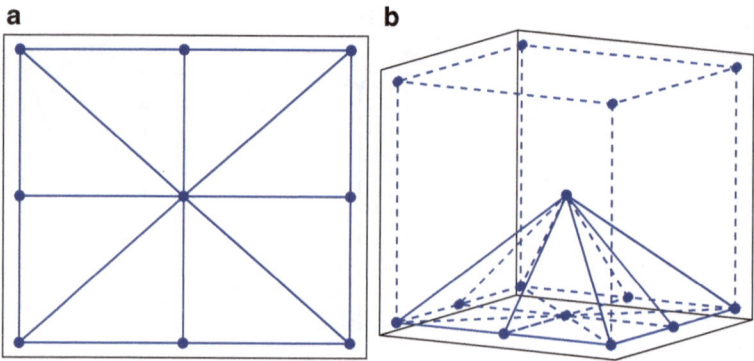

Fig. 1.10 Examples of adaptive triangulation of hyper-rectangles: (**a**) $n = 2$, (**b**) $n = 3$

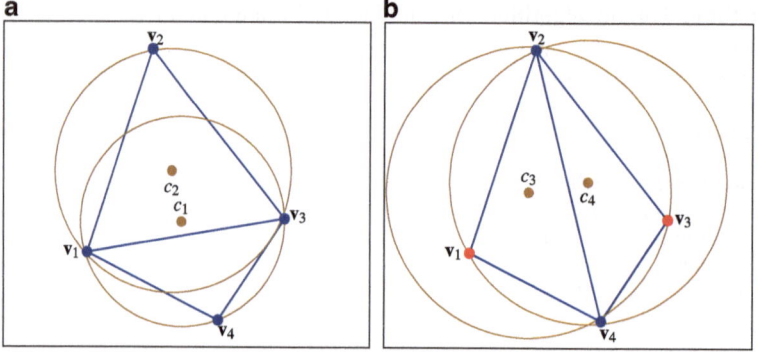

Fig. 1.11 Examples of triangulations: (**a**) Delaunay, (**b**) non-Delaunay

hyper-rectangle $\mathbb{D}$, as well as its geometric center and the centers of gravity of all its faces from dimension 1 to $n - 1$ (3^n points).

Another method of a general face-to-face vertex triangulation is well-known Delaunay triangulation. Delaunay triangulation, named after B. Delaunay work on this topic [20], is widely used in many areas of science and engineering. The fundamental property ("Delaunay condition") of the Delaunay triangulation is often called the empty circumcircle criterion. For a set of points in the Euclidean plane a Delaunay triangulation of these points ensures that the circumcircle associated with a triangle (two dimensional simplex) contains no other point in its interior. Fulfilling the empty circumcircle criterion Delaunay triangulation maximizes the minimum angle of all the triangles. Often Delaunay triangles are said to be "well shaped" because triangles with larger internal angles are selected instead of ones with smaller internal angles.

Delaunay triangulation property is demonstrated in Fig. 1.11. The circumcircles with the centers $\mathbf{c}_1$ and $\mathbf{c}_2$ (see Fig. 1.11a) associated with triangles $[\mathbf{v}_1, \mathbf{v}_3, \mathbf{v}_4]$ and $[\mathbf{v}_1, \mathbf{v}_2, \mathbf{v}_3]$ do not contain points in its interior. On the contrary, the circumcircles

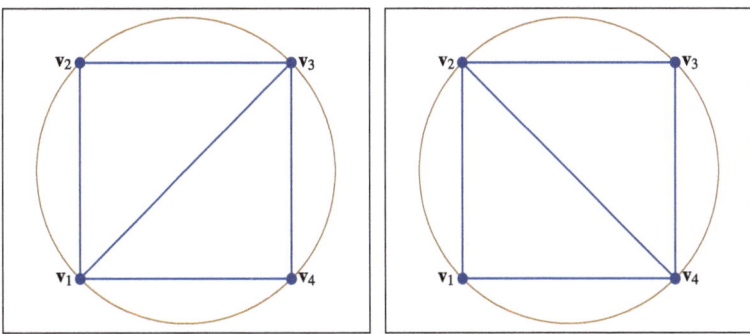

Fig. 1.12 Example of non-unique Delaunay triangulation

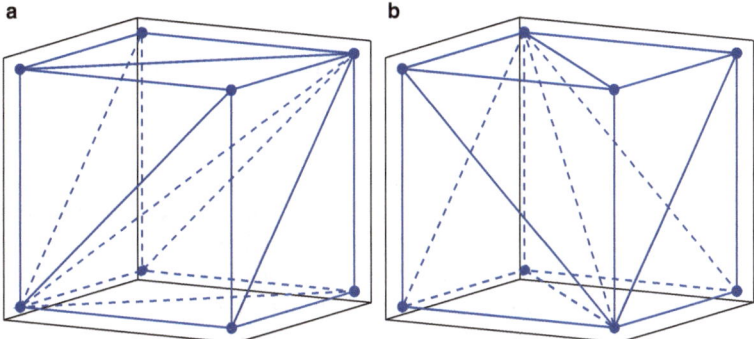

Fig. 1.13 Comparison of triangulation of three-dimensional hyper-rectangle: (**a**) combinatorial triangulation, (**b**) Delaunay triangulation

with the centers c_3 and c_4 (see Fig. 1.11b) associated with triangles $[v_2, v_3, v_4]$ and $[v_1, v_2, v_4]$ are not empty: they contain vertices v_1 and v_3 in their interiors accordingly. The minimal internal angle is maximized by replacing the edge $[v_2, v_4]$ with the edge $[v_1, v_3]$ and therefore the triangulation presented in Fig. 1.11a is a Delaunay triangulation.

While the Delaunay property is well defined, the topology of the triangulation is not unique. In two-dimensional case, such a situation arises when four or more points lie on the same circle. For example the vertices of a square have a non-unique Delaunay triangulation (see Fig. 1.12). Each of the two possible triangulations that subdivide the square into two triangles satisfies the Delaunay condition, i.e., the requirement that the circumcircles of all triangles have empty interiors.

By considering circumscribed spheres, the properties of Delaunay triangulation extend to three and higher dimensions. Figure 1.13 shows a comparison of combinatorial and Delaunay triangulation of three-dimensional hyper-rectangle. The complexity of the Delaunay triangulation increases exponentially in the dimension. Its worst-case complexity is bounded precisely by the theorem, known as the Upper Bound Theorem:

Table 1.1 The number of simplices in triangulations of n-dimensional hyper-rectangle

Triangulation	$n = 2$	$n = 3$	$n = 4$	$n = 5$	$n = 6$	$n = 7$	$n = 8$	$n = 9$	$n = 10$
Combinatorial	2	6	24	120	720	5040	40320	362880	3628800
Delaunay	2	6	22	108	618	4217	33280	273461	2791384

Theorem 1.1 (McMullen [85]). *The number of simplices in the Delaunay triangulation of N points in dimension n is at most $O(N^{\lceil \frac{n}{2} \rceil})$.*

The number of simplices in combinatorial and Delaunay triangulations of n-dimensional ($n = 2, \ldots, 10$) hyper-rectangles are compared in Table 1.1. The numbers of simplices in Delaunay triangulations are smaller; however both numbers increase exponentially in the dimension.

1.4   Covering of Feasible Region Defined by Linear Constraints

Compared to the use of rectangular partitions, simplicial partitions is convenient when the feasible regions is a polytope. Optimization problems with linear constraints are examples where feasible regions are polytopes which can be vertex triangulated. In such a case the constraints may be managed by the initial covering and do not require any consideration later on during the search.

For example, let the optimization problem be defined as

$$\min \ f(\mathbf{x}),$$
$$\text{s.t. } 0 \leq x_i \leq 1,$$
$$x_1 + x_2 \leq 1,$$
$$x_2 - x_3 \leq 0.$$

In this case five of eight vertices of the unit cube defined by $0 \leq x_i \leq 1$ are vertices of the feasible region (see Fig. 1.14):

$$\left\{ \begin{array}{l} (x_1 = 0, x_2 = 0, x_3 = 0) \\ (x_1 = 0, x_2 = 0, x_3 = 1) \\ \cancel{(x_1 = 0, x_2 = 1, x_3 = 0)} \\ (x_1 = 0, x_2 = 1, x_3 = 1) \\ (x_1 = 1, x_2 = 0, x_3 = 0) \\ (x_1 = 1, x_2 = 0, x_3 = 1) \\ \cancel{(x_1 = 1, x_2 = 1, x_3 = 0)} \\ \cancel{(x_1 = 1, x_2 = 1, x_3 = 1)} \end{array} \right\},$$

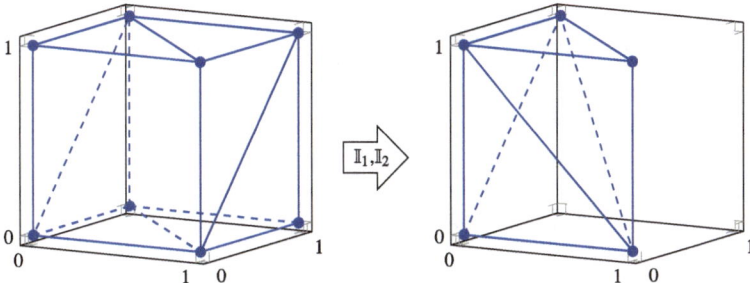

Fig. 1.14 Illustration of covering of feasible region defined by linear inequality constraints

which can be triangulated by two simplices

$$\mathbb{I}_1 = [(0,0,0), (0,0,1), (0,1,1), (1,0,0)],$$
$$\mathbb{I}_2 = [(0,0,1), (0,1,1), (1,0,0), (1,0,1)].$$

Consider another example:

$$\min \ f(\mathbf{x}),$$
$$\text{s.t. } 0 \leq x_i \leq 1,$$
$$x_1 + x_2 \leq 1,$$
$$-x_2 + x_3 \leq 0.$$

In this case four of eight vertices of the unit cube form the feasible region (see Fig. 1.15):

$$\begin{Bmatrix} (x_1 = 0, x_2 = 0, x_3 = 0) \\ \cancel{(x_1 = 0, x_2 = 0, x_3 = 1)} \\ (x_1 = 0, x_2 = 1, x_3 = 0) \\ (x_1 = 0, x_2 = 1, x_3 = 1) \\ (x_1 = 1, x_2 = 0, x_3 = 0) \\ \cancel{(x_1 = 1, x_2 = 0, x_3 = 1)} \\ \cancel{(x_1 = 1, x_2 = 1, x_3 = 0)} \\ \cancel{(x_1 = 1, x_2 = 1, x_3 = 1)} \end{Bmatrix}$$

which is a simplex

$$\mathbb{I} = [(0,0,0), (0,1,0), (0,1,1), (1,0,0)].$$

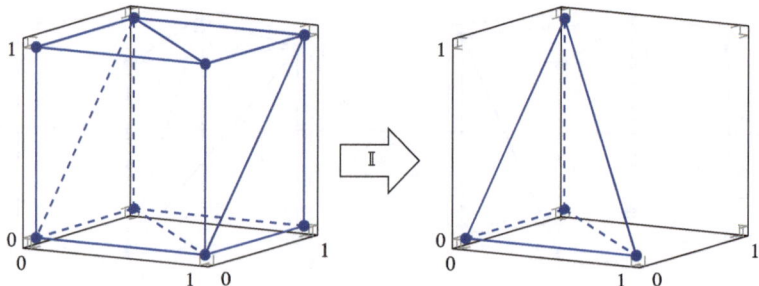

Fig. 1.15 Illustration of simplicial feasible region defined by linear inequality constraints

In a general case, Delaunay triangulation may be used to triangulate a feasible region defined by linear constraints.

If a function is invariant to exchanging of the variables x_i and x_j, we will call it symmetric over the hyper-plane $x_i = x_j$. If such symmetries exist, there are equivalent subregions with equivalent minimizers. The search space can be restricted to avoid equivalent subregions in the search space, and to find only one of the equivalent global minimizers [154, 155]. This may be ensured by restricting the sequence of values of exchangeable variables by setting linear constraints: $x_i \geq x_j$. The resulting constrained search space may be covered by simplices, for example face-to-face vertex triangulated. The search space and the numbers of local and global minimizers may be reduced by avoiding search over equivalent subregions.

If a function is invariant to exchanging all variables, the cubic search space $a \leq x_i \leq b$, $i = 1, \ldots, n$ restricted by the constraints $x_1 \geq x_2 \geq \cdots \geq x_n$ is a simplex and it is $n!$ times smaller than the original cubic search space. Therefore such a simplex may be used as a search space reducing the hyper-volume by $n!$ times and the numbers of minimizers by the similar number of times as well.

For example, the well-known two-dimensional Shubert test function is as follows:

$$f(\mathbf{x}) = \left(\sum_{j=1}^{5} j \cos\left((j+1)x_1 + j\right) \right) \left(\sum_{j=1}^{5} j \cos\left((j+1)x_2 + j\right) \right).$$

The feasible region of the corresponding test problem is $\mathbb{D} = \{\mathbf{x} \in \mathbb{R}^2 : -10 \leq x_i \leq 10, i = 1, \ldots, n\}$. The problem has 18 global minimizers and many local minima.

One can see that the function is invariant to exchange of the variables. In other words, the function is symmetric over the line $x_1 = x_2$. This is also evident in Fig. 1.16b where contour lines of the Shubert test problem are shown as well as simplices partitioned using the DISIMPL-V algorithm. Constraints may be set to avoid equivalent subregions: $x_1 \geq x_2$. The resulting search space is a simplex $[(-10, -10), (10, -10), (10, 10)]$. The situation after termination of later introduced

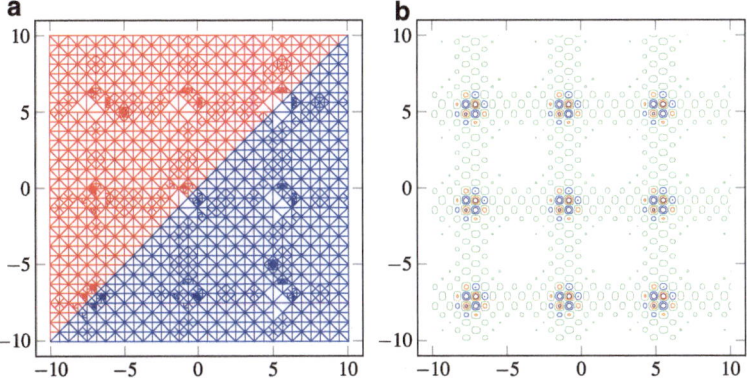

Fig. 1.16 Shubert test problem: (**a**) simplicial partitioning, (**b**) contour lines

Fig. 1.17 Illustration of three-dimensional simplicial feasible region resulted by avoiding equivalent subregions of symmetric function

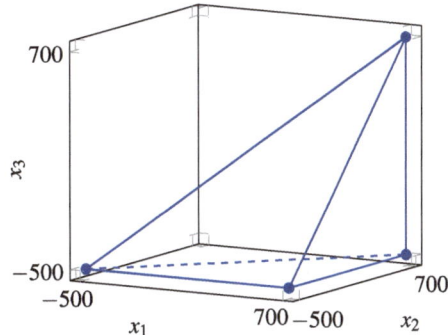

DISIMPL-V algorithm after 30 iterations is presented in Fig. 1.16a: 4509 function evaluations were performed in square search space and almost two times less (2275) in the reduced simplicial search space. The search space and the numbers of local and global minimizers are halved. There are 9 global minimizers in the resulting search space and the other 9 can be reconstructed exchanging the values of variables.

Let us consider another example. The optimization problem is defined as

$$f(\mathbf{x}) = \sum_{i=1}^{n} \frac{x_i^2}{4000} - \prod_{i=1}^{n} \cos(x_i) + 1, \quad \mathbb{D} = [-500, 700]^n.$$

The objective function is symmetric over hyper-planes $x_i = x_j$. Constraints may be set to avoid equivalent subregions: $x_1 \geq x_2 \geq \cdots \geq x_n$. The resulting simplicial search space is

$$\mathbb{I} - [(-500, -500, \ldots, -500), (700, -500, \ldots, -500), \ldots, (700, 700, \ldots, 700)].$$

In the case of a three-dimensional ($n = 3$) problem the resulting simplicial feasible region is illustrated in Fig. 1.17. The search space and the numbers of local and global minimizers are reduced $n!$ times with respect to the original feasible region.

Chapter 2
Lipschitz Optimization with Different Bounds over Simplices

2.1 Lipschitz Optimization

Lipschitz optimization is one of the most deeply investigated subjects of global optimization. It is based on the assumption that the real-valued objective function $f(\mathbf{x})$ and sometimes its first derivative $f'(\mathbf{x})$ are Lipschitz-continuous functions on $\mathbb{D}$, i.e.,

$$|f(\mathbf{x}) - f(\mathbf{y})| \leq L \|\mathbf{x} - \mathbf{y}\|, \ \forall \mathbf{x}, \mathbf{y} \in \mathbb{D}, \quad 0 < L < \infty, \tag{2.1}$$

or

$$\left|f'(\mathbf{x}) - f'(\mathbf{y})\right| \leq K \|\mathbf{x} - \mathbf{y}\|, \ \forall \mathbf{x}, \mathbf{y} \in \mathbb{D}, \quad 0 < K < \infty, \tag{2.2}$$

where $f'(\mathbf{x})$, $f'(\mathbf{y})$ are the directional derivatives calculated at the points $\mathbf{x}, \mathbf{y}$ in the direction $y - x$, L, K are Lipschitz constants, $\mathbb{D} \subset \mathbb{R}^n$ is compact, and $\|\cdot\|$ denotes a norm.

The advantages and disadvantages of Lipschitz global optimization methods are discussed in [49,55,110,134]. These methods are deterministic and provide the same result each time unlike stochastic methods; there is no need to run the algorithm multiple times. In contrast to heuristic methods there is no need for parameter tuning. The algorithms can provide bounds on how far they are from the optimum function value, and hence can use stopping criteria that are more meaningful than a simple iteration limit. The methods guarantee to find an approximation of the solution to a specified accuracy within finite time, as opposed to direct search methods or stochastic methods. They may be used in situation when an analytical description of the objective function is not available (but an estimate of Lipschitz constant is known), as opposed to interval optimization methods.

The main disadvantage of Lipschitz methods is the requirement to provide the Lipschitz constant of the objective function. One of the main questions to be considered in Lipschitz optimization is: how can the Lipschitz constant be obtained?

R. Paulavičius and J. Žilinskas, *Simplicial Global Optimization*,
SpringerBriefs in Optimization, DOI 10.1007/978-1-4614-9093-7_2,
© Remigijus Paulavičius, Julius Žilinskas 2014

There are at least four approaches to obtain an estimate of the Lipschitz constant L [125]:

1. It can be given a priori [3, 59, 89, 112]. This case is very important from the theoretical viewpoint but is not frequently encountered in practice.
2. Its adaptive global estimate over the whole domain can be used [59, 71, 109, 134].
3. Local Lipschitz constants can be estimated adaptively [70, 79, 116, 117, 124, 129, 134];
4. Its estimates can be chosen from a set of possible values [34, 64, 123, 124].

Similar to the Lipschitz constant of the objective function, there are several ways to estimate the Lipschitz constant K of the first derivative. There exist algorithms using an a priori given estimate of K [9, 119], its adaptive estimates [41, 119], adaptive estimates of local Lipschitz constants [27, 44, 118–120], and algorithms working with a number of Lipschitz constants chosen from a set of possible values varying from zero to infinity [72, 73].

There also exist techniques working with Lipschitz multi-extremal constraints [121, 127, 128, 134].

The Lipschitz optimization algorithms investigated in this book can be assigned to the first, second, and fourth approaches to obtain estimate of L. In this chapter we assume that the Lipschitz constant L is known in advance.

The value of the Lipschitz constant in inequality (2.1) depends on the used q-norm

$$\begin{cases} \|\mathbf{x}\|_q = \left(\sum_{i=1}^n |x_i|^q\right)^{1/q}, & 1 \le q < \infty, \mathbf{x} \in \mathbb{R}^n, \\ \|\mathbf{x}\|_\infty = \max\{|x_1|, |x_2|, \dots, |x_n|\}. \end{cases}$$

In Lipschitz optimization the Euclidean norm ($q = 2$) is used most often, but other norms can also be considered [96, 97]. Let us formulate a proposition in which we present the dependence between the Lipschitz constant and the used norm.

Proposition 2.1. *Let $\mathbb{D} \subset \mathbb{R}^n$ be a convex bounded closed set, and let $f(\mathbf{x}) : \mathbb{D} \to \mathbb{R}$ be a continuously differentiable function on an open set containing $\mathbb{D}$. Then $f(\mathbf{x})$ is a Lipschitz function and for $\forall \mathbf{x}, \mathbf{y} \in \mathbb{D}$ the following inequality holds:*

$$|f(\mathbf{x}) - f(\mathbf{y})| \le L_p \|\mathbf{x} - \mathbf{y}\|_q, \tag{2.3}$$

where $L_p = \max\{\|\nabla f(\mathbf{x})\|_p : \mathbf{x} \in \mathbb{D}\}$ is the Lipschitz constant, $\nabla f(\mathbf{x}) = (\frac{\partial f}{\partial x_1}, \dots, \frac{\partial f}{\partial x_n})$ is the gradient of the function $f(\mathbf{x})$, and $\frac{1}{p} + \frac{1}{q} = 1, 1 \le p, q \le \infty$.

Proof. For every pair $\mathbf{x}, \mathbf{y} \in \mathbb{D}$, there exists $\mathbf{z} = \mathbf{x} + \lambda(\mathbf{y} - \mathbf{x}) \in \mathbb{D}, 0 < \lambda < 1$, such that

$$|f(\mathbf{y}) - f(\mathbf{x})| = |\nabla f(\mathbf{z}) \cdot (\mathbf{y} - \mathbf{x})|,$$

(by the mean-value theorem)

$$|f(\mathbf{y}) - f(\mathbf{x})| \le \sum_{i=1}^{n} \left| \frac{\partial f}{\partial z_i}(y_i - x_i) \right| \le \left(\sum_{i=1}^{n} \left| \frac{\partial f}{\partial z_i} \right|^p \right)^{1/p} \left(\sum_{i=1}^{n} |y_i - x_i|^q \right)^{1/q},$$

where $1/p + 1/q = 1$, $1 \le p, q \le \infty$ (using Hölder's inequality).
Since

$$\|\nabla f(\mathbf{z})\|_p = \left(\sum_{i=1}^{n} \left| \frac{\partial f}{\partial z_i} \right|^p \right)^{1/p}, \quad \|\mathbf{x} - \mathbf{y}\|_q = \left(\sum_{i=1}^{n} |x_i - y_i|^q \right)^{1/q},$$

therefore

$$|f(\mathbf{x}) - f(\mathbf{y})| \le \|\nabla f(\mathbf{z})\|_p \|\mathbf{x} - \mathbf{y}\|_q \le \max\left\{ \|\nabla f(\mathbf{z})\|_p : \mathbf{z} \in \mathbb{D} \right\} \|\mathbf{x} - \mathbf{y}\|_q.$$

Denoting $L_p = \max\left\{ \|\nabla f(\mathbf{z})\|_p : \mathbf{z} \in \mathbb{D} \right\}$ we get (2.3).

In conclusion, the supremum of the p norm of the gradient of the function is the Lipschitz constant for the q norm in the Lipschitz condition.

In this chapter we assume that the Lipschitz constants are known. For the experimental investigation of the Lipschitz bounds and algorithms, the constants have been estimated following the procedure described in [49]. A grid search algorithm of 1000^n points for ($n = 2, 3$) and 100^n points for ($n = 4, 5, 6$) dimensional test problems (which are given together with their derivatives in Appendix A) has been used and the obtained estimates are thus close to the actual Lipschitz constants.

The estimated Lipschitz constants' values $L_p : p = 1, 2, \infty$ and the optimal vectors $\mathbf{z} = (z_1, \ldots, z_n) \in \mathbb{D}$ where the maximal p-norm was obtained are shown in Tables 2.1 and 2.2. In the tables the symbol # means "Test Problem number."

The estimates of the Euclidean Lipschitz constants L_2 were compared to those given in [49] which are repeated in Table 2.1. Most of the estimates coincide. However, the estimates of the Lipschitz constants of the test problems 14 and 15 are quite different. We believe that our estimates are more correct. In the case of the test problem 14, the values of derivatives at the feasible point $(1, 1, 0) \in \mathbb{D} = [0, 1]^3$ are equal to 200, and therefore from (2.3)

$$L_2 = \|\nabla f(1, 1, 0)\|_2 = \sqrt{200^2 + 200^2 + 200^2} \approx 346.41.$$

2.2 Classical Lipschitz Bounds

The most studied case of problem (1.1) is the bound constrained univariate one ($n = 1$), for which numerous algorithms have been proposed, compared, and theoretically investigated. An excellent comprehensive survey is contained in [49]. In this book, we are mainly interested in the multivariate case ($n \ge 2$).

Table 2.1 Lipschitz constants values for $(n = 2, 3)$ dimensional test problems using $\infty, 2, 1$-norms

#	Lipschitz constants				Optimal vector		
	L_1	L_2	$L_2[49]$	L_∞	z_1	z_2	z_3
1	50.27	50.27	50.2665	50.27	1.000	1.000	–
2	7.98	6.32	6.3183	6.00	1.000	0.380	–
3	141.34	112.44	112.44	107.09	0.000	0.000	–
4	2.00	2.00	2	2.00	1.000	1.000	–
5	2.00	1.42	$\sqrt{2}$	1.00	0.000	0.000	–
6	1284.80	1059.59	1059.59		4.000	4.000	–
				1034.00	−2.000	4.000	–
7	14708.00	12781.70	12781.7	12608.00	−3.000	−1.500	–
8	122.00	86.32	86.3134	63.00	−2.500	4.500	–
9	2639040.00	2225890.00	2225892	2177520.00	−2.000	2.000	–
10	24.00	17.03	17.034	13.04	4.000	−3.000	–
11	64.14	47.55	47.426	42.16	1.000	2.000	–
12	80.40	56.85	56.852	40.40	1.000	1.000	–
13		993.72	988.82	993.72	0.010	0.458	–
	995.29				0.010	0.020	–
14	600.00	346.41	244.95	200.00	1.000	1.000	0.000
15		18.33	42.626	18.32	0.091	0.556	0.990
	21.36				0.202	0.444	0.970
16		988.79	971.59	988.79	0.010	0.130	0.250
	991.05				0.010	0.020	0.210
17	33.52	19.39	19.39		4.000	3.528	4.000
				14.80	1.177	4.000	4.000
18	4.77	2.92	2.919	2.38	1.000	1.000	1.000
19	224.00	130.00	130.0	80.00	2.000	−2.000	2.000
20	33616.00	22267.90	–	16808.00	−3.000	−3.000	−3.000

In Lipschitz optimization the lower bound for $f(\mathbf{x})$ over a subregion $\mathbb{I} \subseteq \mathbb{D}$ is evaluated by exploiting the Lipschitz condition (2.3). It follows from (2.3) that for all $\mathbf{x}, \mathbf{y} \in \mathbb{I}$

$$f(\mathbf{x}) \geq f(\mathbf{y}) - L_p \|\mathbf{x} - \mathbf{y}\|_q.$$

If $\mathbf{y} \in \mathbb{I}$ is fixed, then the concave function

$$F_{\mathbf{y}}^q(\mathbf{x}) = f(\mathbf{y}) - L_p \|\mathbf{x} - \mathbf{y}\|_q \qquad (2.4)$$

underestimates $f(\mathbf{x})$ over $\mathbb{I}$. The superscript q indicates the norm used for the bound calculation.

Thus an algorithm for Lipschitz optimization may be built. Start from an arbitrary point $\mathbf{x}_1 \in \mathbb{I}$ and define the first underestimating function by

$$F_1^q(\mathbf{x}) = f(\mathbf{x}_1) - L_p \|\mathbf{x} - \mathbf{x}_1\|_q. \qquad (2.5)$$

Table 2.2 Lipschitz constants values for ($n \geq 4$) test functions using ∞, 2, 1-norms

#	Lipschitz constants			Optimal vector					
	L_1	L_2	L_∞	z_1	z_2	z_3	z_4	z_5	z_6
21	1370.2	1196.4	1195.5	−10.00	−10.00	−10.00	−9.58	–	–
22	108720	60402	36026	4.00	−4.00	−4.00	−4.00	–	–
23	204.1	102.4	56.1	10.00	0.00	0.00	0.00	–	–
24	300.1	151.5	86.1	0.00	0.00	10.00	0.00	–	–
25	408.2	204.5	110.8	0.00	0.00	10.00	0.00	–	–
26	600.0	313.7	200.0	10.00	10.00	10.00	10.00	–	–
27	92216	48252		−4.00	−4.00	5.00	5.00	–	–
			29270	5.00	5.00	−4.00	−4.00	–	–
28	26.1	14.4	8.3	−10.00	−10.00	−10.00	−10.00	–	–
29		370.7	369.7	−5.00	−5.00	−5.00	−5.00	−5.00	–
	476.7			−5.00	−5.00	−5.00	−4.92	−4.58	–
30	264385	129425	66032	5.00	−5.00	−5.00	−5.00	−5.00	–
31	34.4	16.6	8.3	−10.00	−10.00	−10.00	−10.00	−10.00	–
32		370.9	369.7	−5.00	−5.00	−5.00	−5.00	−5.00	−4.59
	488.7			−5.00	−5.00	−5.00	−5.00	−4.92	−4.58
33	546547	240918	109238	6.00	−6.00	−6.00	−6.00	−6.00	−6.00

Then by choosing the "most promising" point $x_2 \in \mathbb{I}$ (with the minimal lower-bounding function $F_1^q(\mathbf{x})$ value)

$$\mathbf{x}_2 = \arg\min_{\mathbf{x}\in\mathbb{I}} F_1^q(\mathbf{x}),$$

we obtain another concave underestimating function $f(\mathbf{x}_2) - L_p\|\mathbf{x} - \mathbf{x}_2\|_q$, and the best underestimator of $f(\mathbf{x})$ on $\mathbb{I}$, given the information obtained so far, is

$$F_2^q(\mathbf{x}) = \max_{i=1,2}\{f(\mathbf{x}_i) - L_p\|\mathbf{x} - \mathbf{x}_i\|_q\}.$$

By analogy, in step k the underestimating function $F_k^q(\mathbf{x})$ is

$$F_k^q(\mathbf{x}) = \max_{i=1,\dots,k}\left\{f(\mathbf{x}_i) - L_p\|\mathbf{x} - \mathbf{x}_i\|_q\right\}, \qquad (2.6)$$

and the next iteration point $\mathbf{x}_{k+1}$ is given by

$$\mathbf{x}_{k+1} = \arg\min_{\mathbf{x}\in\mathbb{I}} F_k^q(\mathbf{x}). \qquad (2.7)$$

The lower bound based on (2.6) and (2.7) calculations will be referred as the φ type (or Piyavskii [112] type) bound

$$\varphi^q(\mathbb{I}) = \min_{\mathbf{x}\in\mathbb{I}} \max_{\mathbf{x}_i\in\mathbb{T}} F_{\mathbf{x}_i}^q(\mathbf{x}), \qquad (2.8)$$

where $\mathbb{T}$ is a finite set of distinct points in $\mathbb{I}$.

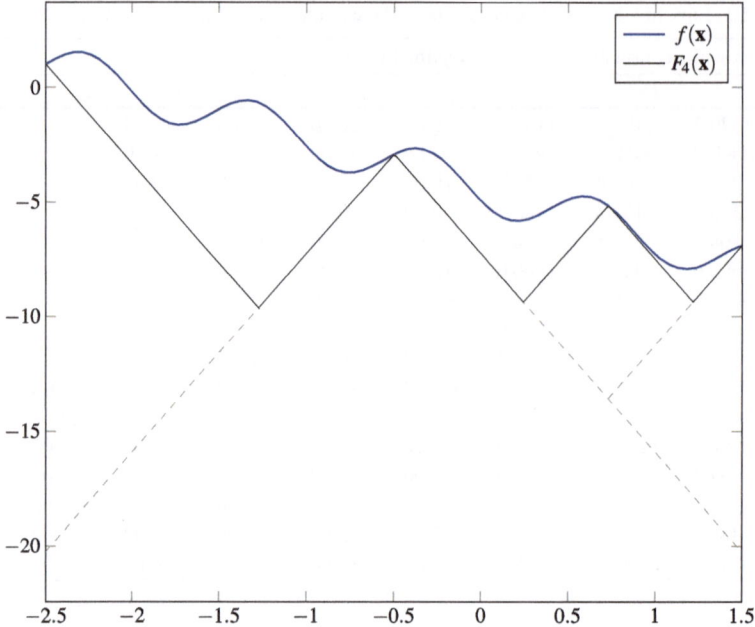

Fig. 2.1 Visualization of Piyavskii type lower bounding function $F_k(\mathbf{x})$ in univariate case

In the univariate case ($n = 1$) all norms are equal, therefore q-norm superscript is not used. The function $\max_{\mathbf{x}_i \in \mathbb{T}} F_{\mathbf{x}_i}(\mathbf{x})$ is piecewise linear (see Fig. 2.1), and φ can be determined in a simple straightforward way [49]. Therefore, many univariate algorithms use the bound φ, where the set $\mathbb{T}$ is suitably updated in an iterative way. The most often studied of these methods is due to Piyavskii [111, 112]. It was independently rediscovered by Shubert [130]. Figure 2.1 visualizes univariate Piyavskii method situation after $k = 4$ steps, where

$$f(x) = -\frac{13}{6}x + \sin\left(\frac{13}{4}(2x + 5)\right) - \frac{53}{12}, \quad \mathbb{D} = [-2.5, 1.5], \quad L = 8.66. \quad (2.9)$$

When $\mathbb{I}$ is a two-dimensional rectangle in $\mathbb{R}^2$, φ type bound with the Euclidean norm ($q = 2$) can still be evaluated by geometric arguments which take into account the conical shape of the lower bounding function [89]. For ($n > 2$), however, problem (2.8) constitutes a difficult optimization problem. Convergent deterministic methods for solving the multivariate unconstrained problem (1.1) fall into three main classes.

First, multivariate Lipschitz optimization can be reduced to the univariate case. Following this idea, a nested optimization scheme was proposed by Piyavskii [111, 112]. Another scheme based on filling the feasible space by Peano curve was studied by Butz [11] and Strongin [132].

The second class contains direct extensions of Piyavskii's method [111, 112] to the multivariate case. Various modifications using the Euclidean norm: Piyavskii [112], Mladineo [89, 90], Jaumard et al. [61], or other norms or close approximations: Mayne and Polak [84], Wood [141, 142], Zhang et al. [144] have been proposed. Note that when the Euclidean norm is used in the multivariate case, the lower bounding functions are envelopes of circular cones with parallel symmetry axes. A problem of finding the minimum (2.8) of such a bounding function becomes a difficult global optimization problem involving a system of quadratic and linear equations. Most of these approaches are quite ingenious from a theoretical viewpoint, but the inherent difficulty of subproblems has limited practical applicability to dimension $n > 2$ in general unconstrained problems. For an excellent survey and numerical tests, see [49]. Most of these algorithms can be improved by interpreting them as branch-and-bound methods.

The third class contains many simplicial and rectangular branch-and-bound techniques, but, in general, considerably weaker bounds: Galperin [37, 38], Pinter [106–108], Meewella and Mayne [86, 87], Gourdin et al. [45]. These algorithms fit into the general framework proposed by Horst [53], Horst and Tuy [58, 59]. These algorithms differ in the selection rules, the ways how branching is performed and bounds are computed.

Lipschitz branch-and-bound algorithms usually use considerably weaker bounds. Most often, weaker bounds belong to the following two simple families μ_1 and μ_2. Let

$$\delta_q(\mathbb{I}) = \max \left\{ \|\mathbf{x} - \mathbf{y}\|_q : \mathbf{x}, \mathbf{y} \in \mathbb{I} \right\}$$

denote the diameter of $\mathbb{I}$. For example, if $\mathbb{I} = \{\mathbf{x} \in \mathbb{R}^n : a_i \leq x_i \leq b_i\}$ is an n-rectangle, then $\delta_q(\mathbb{I}) = \|\mathbf{b} - \mathbf{a}\|_q$, and if $\mathbb{I}$ is an n-simplex, then the diameter $\delta_q(\mathbb{I})$ is the length of its longest edge.

A simpler lower bound can be derived from (2.4):

$$\mu_1^q(\mathbb{I}) = \max_{\mathbf{y} \in \mathbb{T}} f(\mathbf{y}) - L_p \delta_q(\mathbb{I}), \tag{2.10}$$

where $\mathbb{T} \subset \mathbb{I}$ is a finite sample of points in $\mathbb{I}$, where the function values of f have been evaluated. If $\mathbb{I}$ is a rectangle or a simplex, the set $\mathbb{T}$ often coincides with (or is a subset of) the vertex set $\mathbb{V}(\mathbb{I})$. Note, however, that it might be worthwhile to take into account interior points. The middle point $\mathbf{m} = (\mathbf{a} + \mathbf{b})/2$ of an n-rectangle $\mathbf{a} \leq \mathbf{x} \leq \mathbf{b}$ yields the bound

$$\mu_m^q(\mathbb{I}) = f(\mathbf{m}) - L_p \delta_q(\mathbb{I})/2.$$

A tighter but computationally more expensive than (2.10) bound is

$$\mu_2^q(\mathbb{I}) = \max_{\mathbf{y} \in \mathbb{T}} \left\{ f(\mathbf{y}) - L_p \max_{\mathbf{z} \in \mathbb{V}(\mathbb{I})} \|\mathbf{y} - \mathbf{z}\|_q \right\}. \tag{2.11}$$

This situation suggests that searching for new lower bounds, which are tighter than (2.10) and (2.11), but computationally less expensive than (2.8), might be worthwhile. In the next sections, lower bounds over hyper-rectangles and simplices are discussed. These improved bounds are stronger than (2.10) and (2.11), but they are usually still computationally cheap.

2.3   Impact of Norms on Lipschitz Bounds

In the one-dimensional case, all norms are equal. But in the multidimensional case evaluated bounds depend on the used norm [96, 97]. In this section, we investigate how the Lipschitz bounds are influenced by the used norms and the corresponding Lipschitz constants when μ_2^q bound (2.11) is used.

Let us investigate simplicial partitioning of a hyper-cubic feasible region. In two-dimensional case such a feasible region $\mathbb{D} = [0, 1]^2$ is a unit square, which by face-to-face vertex triangulation is divided into two triangles (two-dimensional simplices) $\mathbb{I}_1 = [\mathbf{v}_1, \mathbf{v}_3, \mathbf{v}_4]$ and $\mathbb{I}_2 = [\mathbf{v}_1, \mathbf{v}_2, \mathbf{v}_4]$ (see Fig. 2.2).

Let's freely choose one of the simplex, say $\mathbb{I}_1$, and calculate the μ_2^q type lower bound:

$$\mu_2^q(\mathbb{I}_1) = \max_{\mathbf{v}_i \in \mathbb{V}(\mathbb{I}_1)} \left\{ f(\mathbf{v}_i) - L_p \max_{\mathbf{x} \in \mathbb{V}(\mathbb{I}_1)} \|\mathbf{x} - \mathbf{v}_i\|_q \right\}. \tag{2.12}$$

For $\mathbf{v}_1$ (or $\mathbf{v}_4$) the maximal distance

$$\max_{\mathbf{x} \in \mathbb{V}(\mathbb{I}_1)} \|\mathbf{x} - \mathbf{v}_1\|_q$$

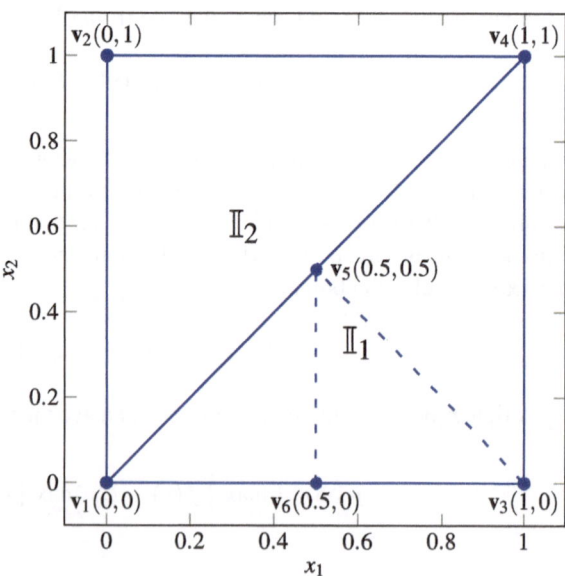

Fig. 2.2 Covering of square by simplices (*triangles*)

is between vertices $\mathbf{v}_1$ and $\mathbf{v}_4$, which depending on the used q-norm is equal to

$$\|\mathbf{v}_1 - \mathbf{v}_4\|_1 = (|0 - 1| + |0 - 1|) = 2,$$

$$\|\mathbf{v}_1 - \mathbf{v}_4\|_q = (|0 - 1|^q + |0 - 1|^q)^{\frac{1}{q}} = 2^{\frac{1}{q}}, \quad 1 < q < \infty,$$

$$\|\mathbf{v}_1 - \mathbf{v}_4\|_\infty = \max\{|0 - 1|, |0 - 1|\} = 1.$$

From (2.3), the product

$$L_p \max_{\mathbf{x} \in \mathbb{V}(\mathbb{I}_1)} \|\mathbf{x} - \mathbf{v}_i\|_q$$

depending on the used norm and the corresponding Lipschitz constant is equal to

$$L_\infty \|\mathbf{v}_1 - \mathbf{v}_4\|_1 = \max\{|f'_{x_1}(\mathbf{z})|, |f'_{x_2}(\mathbf{z})|\} \cdot 2,$$

$$L_p \|\mathbf{v}_1 - \mathbf{v}_4\|_q = (|f'_{x_1}(\mathbf{z})|^p + |f'_{x_2}(\mathbf{z})|^p)^{\frac{1}{p}} \cdot 2^{\frac{1}{q}}, \quad \frac{1}{p} + \frac{1}{q} = 1, \ 1 < p, q < \infty,$$

$$L_1 \|\mathbf{v}_1 - \mathbf{v}_4\|_\infty = (|f'_{x_1}(\mathbf{z})| + |f'_{x_2}(\mathbf{z})|) \cdot 1,$$

where $\mathbf{z} = (z_1, z_2) \in \mathbb{D}$ is the point where the q-norm of the gradient of the objective function is largest. It can be noticed in Tables 2.1 and 2.2 that the points $\mathbf{z}$ may differ for various norms. However, for simplicity we assume that they are identical as is the case for most of the test problems in Tables 2.1 and 2.2.

Lemma 2.1 ([67]). *Let $p > 0$ and $a, b \geq 0$. Then*

$$(a + b)^p \leq \max\{1, 2^{p-1}\}(a^p + b^p).$$

Using Lemma 2.1 we get

$$(|f'_{x_1}(\mathbf{z})| + |f'_{x_2}(\mathbf{z})|) \cdot 1 \leq 2^{\frac{p-1}{p}} (|f'_{x_1}(\mathbf{z})|^p + |f'_{x_2}(\mathbf{z})|^p)^{\frac{1}{p}}. \tag{2.13}$$

Expressing $q = \frac{p}{p-1}$ and inserting into (2.13) we get

$$(|f'_{x_1}(\mathbf{z})| + |f'_{x_2}(\mathbf{z})|) \cdot 1 \leq (|f'_{x_1}(\mathbf{z})|^p + |f'_{x_2}(\mathbf{z})|^p)^{\frac{1}{p}} \cdot 2^{\frac{1}{q}}$$

$$\leq \max\{|f'_{x_1}(\mathbf{z})|, |f'_{x_2}(\mathbf{z})|\} \cdot 2,$$

i.e.,

$$L_1 \|\mathbf{v}_1 - \mathbf{v}_4\|_\infty \leq L_p \|\mathbf{v}_1 - \mathbf{v}_4\|_q \leq L_\infty \|\mathbf{v}_1 - \mathbf{v}_4\|_1. \tag{2.14}$$

From (2.14) it follows that for simplex $\mathbb{I}_1$ when the function values at vertices $\mathbf{v}_1$ and $\mathbf{v}_4$ are used, the best bound of μ_2^q type is with the ∞-norm and the corresponding

Lipschitz constant L_1, i.e.,

$$\mu_2^\infty(\mathbf{v}_i) = f(\mathbf{v}_i) - L_1\|\mathbf{v}_1 - \mathbf{v}_4\|_\infty, \tag{2.15}$$

where $i = 1$ or $i = 4$.

Calculating the bound over the simplex $\mathbb{I}_1$ vertex $\mathbf{v}_3$, it is easy to see that the distance to vertex $\mathbf{v}_1$ (or $\mathbf{v}_4$) independently on the used norm is equal to 1:

$$\|\mathbf{v}_3 - \mathbf{v}_1\|_1 = (|1 - 0| + |0 - 0|) = 1,$$

$$\|\mathbf{v}_3 - \mathbf{v}_1\|_q = (|1 - 0|^q + |0 - 0|^q)^{\frac{1}{q}} = 1, \quad 1 < q < \infty,$$

$$\|\mathbf{v}_3 - \mathbf{v}_1\|_\infty = \max\{|1 - 0|, |0 - 0|\} = 1.$$

Therefore when calculating the bound using the function value at vertex $\mathbf{v}_3$, the best bound is with the 1-norm and the corresponding L_∞ Lipschitz constant:

$$\mu_2^1(\mathbf{v}_3) = f(\mathbf{v}_3) - L_\infty\|\mathbf{v}_3 - \mathbf{v}_1\|_1, \tag{2.16}$$

because $L_\infty \leq L_p, 1 \leq p < \infty$.

In summary the best μ_2^q type bound over the simplex $\mathbb{I}_1$ is either with the 1-norm and the corresponding Lipschitz constant L_∞ (2.16) or the ∞-norm and the corresponding Lipschitz constant L_1 (2.15).

It can be seen that for the $\mathbb{I}_2$ simplex and any simplex (isosceles right triangle $[\mathbf{v}_3, \mathbf{v}_4, \mathbf{v}_5], [\mathbf{v}_3, \mathbf{v}_5, \mathbf{v}_6], \ldots$ (see Fig. 2.2)) later obtained using subdivision into two through the middle point of the longest edge, the best μ_2^q type bound are achieved, when either (2.15) or (2.16) is used.

Therefore to get tighter μ_2^q type bounds over two-dimensional simplices, it is useful to combine the 1-norm and the ∞-norm with the corresponding Lipschitz constants L_∞ and L_1:

$$\mu_2^{1,\infty}(\mathbb{I}) = \max_{\mathbf{v}_i \in V(\mathbb{I})} \left\{ f(\mathbf{v}_i) - \min\left\{ L_\infty \max_{\mathbf{x} \in V(\mathbb{I})} \|\mathbf{x} - \mathbf{v}_i\|_1, L_1 \max_{\mathbf{x} \in V(\mathbb{I})} \|\mathbf{x} - \mathbf{v}_i\|_\infty \right\} \right\}. \tag{2.17}$$

We observe that using μ_1^q type bound with the 1-norm and the ∞- norm also tighter bounds are achieved, comparing to the bound when just one of the norms is used:

$$\mu_1^{1,\infty}(\mathbb{I}) = \max_{\mathbf{v}_i \in V(\mathbb{I})} f(\mathbf{v}_i) - \min\{L_\infty \delta_1(\mathbb{I}), L_1 \delta_\infty(\mathbb{I})\}. \tag{2.18}$$

When the feasible region is a rectangle (not a square), after the initial vertex triangulation of the feasible region, the simplices (triangles) are not isosceles, therefore (2.14) is not always correct and proposed bound $\mu_2^{1,\infty}$ (2.17) is not

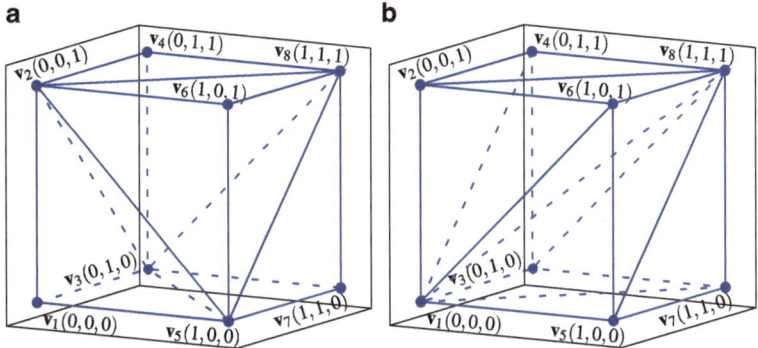

Fig. 2.3 Covering of a unit cube by simplices: (**a**) 5 tetrahedrons, (**b**) combinatorial triangulation with $3! = 6$ simplices

always better than one with a concrete norm. However, experimental investigation in Sect. 2.6 shows that the combined bound $\mu_2^{1,\infty}$ significantly improves optimization results for the rectangular feasible regions as well.

Let us extend our investigation when the feasible region is a (unit) cube $\mathbb{D} = [0, 1]^3$. We investigate two different strategies of initial cube triangulation described in Sect. 1.3: face-to-face vertex triangulation by five simplices (see Fig. 2.3a) and combinatorial triangulation by six simplices (see Fig. 2.3b).

Let's start from the case, where the unit cube is triangulated into five tetrahedrons (see Fig. 2.3a). After such a face-to-face vertex triangulation we get four equal volume simplices and the middle one $[\mathbf{v}_2, \mathbf{v}_3, \mathbf{v}_5, \mathbf{v}_8]$—with two times bigger volume. Let us investigate one of the four similar simplices, say the simplex $[\mathbf{v}_1, \mathbf{v}_2, \mathbf{v}_3, \mathbf{v}_5]$, and evaluate μ_2^q type bound using the function values at all vertices analogously as we did in the two-dimensional case.

When the bound is evaluated using the vertex $\mathbf{v}_1$, the maximal distance

$$\max_{\mathbf{x} \in [\mathbf{v}_2, \mathbf{v}_3, \mathbf{v}_5]} \|\mathbf{x} - \mathbf{v}_1\|_q$$

is equal to the distance to any of other vertices (lets take $\mathbf{v}_2$) and independently on the used norm is equal to 1:

$$\|\mathbf{v}_1 - \mathbf{v}_2\|_1 = (|0 - 0| + |0 - 0| + |0 - 1|) = 1,$$

$$\|\mathbf{v}_1 - \mathbf{v}_2\|_q = (|0 - 0|^q + |0 - 0|^q + |0 - 1|^q)^{\frac{1}{q}} = 1, \quad 1 < q < \infty,$$

$$\|\mathbf{v}_1 - \mathbf{v}_2\|_\infty = \max\{|0 - 0|, |0 - 0|, |0 - 1|\} = 1.$$

Therefore when using the function value at the vertex $\mathbf{v}_1$, the best bound is with the 1-norm and the corresponding Lipschitz constant L_∞:

$$\mu_2^1(\mathbf{v}_1) = f(\mathbf{v}_1) - L_\infty \|\mathbf{v}_1 - \mathbf{v}_2\|_1. \tag{2.19}$$

When evaluating bounds using the remaining vertices $\mathbf{v}_2, \mathbf{v}_3, \mathbf{v}_5$, the maximal distance, independently on the vertices (let's take $\mathbf{v}_2$)

$$\max_{\mathbf{x} \in [\mathbf{v}_1, \mathbf{v}_3, \mathbf{v}_5]} \|\mathbf{x} - \mathbf{v}_2\|_q$$

is between $\mathbf{v}_2$ and $\mathbf{v}_3$ (or between $\mathbf{v}_2$ and $\mathbf{v}_5$), which according to the used norm is equal to:

$$\|\mathbf{v}_2 - \mathbf{v}_3\|_1 = (|0 - 0| + |0 - 1| + |1 - 0|) = 2,$$

$$\|\mathbf{v}_2 - \mathbf{v}_3\|_q = (|0 - 0|^q + |0 - 1|^q + |1 - 0|^q)^{\frac{1}{q}} = 2^{\frac{1}{q}}, \quad 1 < q < \infty,$$

$$\|\mathbf{v}_2 - \mathbf{v}_3\|_\infty = \max\{|0 - 1|, |0 - 0|, |1 - 0|\} = 1.$$

Therefore the following products of the Lipschitz constant and the corresponding norm may be

$$L_\infty \|\mathbf{v}_2 - \mathbf{v}_3\|_1 = \max\{|f'_{x_1}(\mathbf{z})|, |f'_{x_2}(\mathbf{z})|, |f'_{x_3}(\mathbf{z})|\} \cdot 2,$$

$$L_p \|\mathbf{v}_2 - \mathbf{v}_3\|_q = (|f'_{x_1}(\mathbf{z})|^p + |f'_{x_2}(\mathbf{z})|^p + |f'_{x_3}(\mathbf{z})|^p)^{\frac{1}{p}} \cdot 2^{\frac{1}{q}}, 1 < p, q < \infty,$$

$$L_1 \|\mathbf{v}_2 - \mathbf{v}_3\|_\infty = (|f'_{x_1}(\mathbf{z})| + |f'_{x_2}(\mathbf{z})| + |f'_{x_3}(\mathbf{z})|) \cdot 1.$$

By analogy to (2.14), for the three-dimensional case, the inequality

$$(|f'_{x_1}(\mathbf{z})| + |f'_{x_2}(\mathbf{z})| + |f'_{x_3}(\mathbf{z})|) \leq (|f'_{x_1}(\mathbf{z})|^p + |f'_{x_2}(\mathbf{z})|^p + |f'_{x_3}(\mathbf{z})|^p)^{\frac{1}{p}} \cdot 2^{\frac{1}{q}}$$

$$\leq \max\{|f'_{x_1}(\mathbf{z})|, |f'_{x_2}(\mathbf{z})|, |f'_{x_3}(\mathbf{z})|\} \cdot 2$$

$$(2.20)$$

is correct only when one of the partial derivatives $|f'_{x_i}(\mathbf{z})|$, $i = 1, \ldots, 3$ is equal or close to 0. Therefore for three-dimensional case $\mu_2^{1,\infty}$ (2.17) is not always better than μ_2^q with one concrete q-norm.

For the middle simplex $[\mathbf{v}_2, \mathbf{v}_3, \mathbf{v}_5, \mathbf{v}_8]$, the maximum distance from any vertex to other vertices is the same as in the previously investigated simplex $[\mathbf{v}_1, \mathbf{v}_2, \mathbf{v}_3, \mathbf{v}_5]$: the distance between $\mathbf{v}_2$ and $\mathbf{v}_3$. It follows that $\mu_2^{1,\infty}$ is not always the best to this simplex too.

The investigated covering is not known in a general case for $n > 3$. Therefore it is important to investigate the bounds using a general covering, e.g., combinatorial triangulation (see Fig. 2.3b).

We note that in such a covering, vertices $\mathbf{v}_1$ and $\mathbf{v}_8$ belong to all simplices and all simplices are of equal volume. Taking any of the simplices, say $[\mathbf{v}_1, \mathbf{v}_2, \mathbf{v}_4, \mathbf{v}_8]$ we can evaluate μ_2^q bounds. The maximal distance is between vertices $\mathbf{v}_1$ and $\mathbf{v}_8$, according to the used norm

$$\|\mathbf{v}_1 - \mathbf{v}_8\|_1 = (|0 - 1| + |0 - 1| + |0 - 1|) = 3,$$

$$\|\mathbf{v}_1 - \mathbf{v}_8\|_q = (|0 - 1|^q + |0 - 1|^q + |0 - 1|^q)^{\frac{1}{q}} = 3^{\frac{1}{q}}, \quad 1 < q < \infty,$$

$$\|\mathbf{v}_1 - \mathbf{v}_8\|_\infty = \max\{|0 - 1|, |0 - 1|, |0 - 1|\} = 1.$$

Therefore the following products are possibly depending on the used norms and the corresponding Lipschitz constants:

$$L_\infty \|\mathbf{v}_1 - \mathbf{v}_8\|_1 = \max\{|f'_{x_1}(\mathbf{z})|, |f'_{x_2}(\mathbf{z})|, |f'_{x_3}(\mathbf{z})|\} \cdot 3,$$

$$L_p \|\mathbf{v}_1 - \mathbf{v}_8\|_q = (|f'_{x_1}(\mathbf{z})|^p + |f'_{x_2}(\mathbf{z})|^p + |f'_{x_3}(\mathbf{z})|^p)^{\frac{1}{p}} \cdot 3^{\frac{1}{q}}, \quad 1 < p, q < \infty,$$

$$L_1 \|\mathbf{v}_1 - \mathbf{v}_8\|_\infty = (|f'_{x_1}(\mathbf{z})| + |f'_{x_2}(\mathbf{z})| + |f'_{x_3}(\mathbf{z})|) \cdot 1.$$

In such a case, inequalities

$$(|f'_{x_1}(\mathbf{z})| + |f'_{x_2}(\mathbf{z})| + |f'_{x_3}(\mathbf{z})|) \leq (|f'_{x_1}(\mathbf{z})|^p + |f'_{x_2}(\mathbf{z})|^p + |f'_{x_3}(\mathbf{z})|^p)^{\frac{1}{p}} \cdot 3^{\frac{1}{q}}$$

$$\leq \max\{|f'_{x_1}(\mathbf{z})|, |f'_{x_2}(\mathbf{z})|, |f'_{x_3}(\mathbf{z})|\} \cdot 3,$$

i.e.,

$$L_1 \|\mathbf{v}_1 - \mathbf{v}_8\|_\infty \leq L_p \|\mathbf{v}_1 - \mathbf{v}_8\|_q \leq L_\infty \|\mathbf{v}_1 - \mathbf{v}_8\|_1 \tag{2.21}$$

hold. Therefore, the μ_2^q type bound calculated using vertex $\mathbf{v}_1$ (or $\mathbf{v}_8$) is best when the ∞-norm and the corresponding Lipschitz constant is used:

$$f(\mathbf{v}_1) - L_1 \|\mathbf{v}_1 - \mathbf{v}_8\|_\infty. \tag{2.22}$$

Therefore after combinatorial triangulation, all simplices have two vertices, in which μ_2^q type bound with the infinity norm and the corresponding Lipschitz constant should be used.

Let's evaluate bound on the remaining vertices $\mathbf{v}_2$ and $\mathbf{v}_4$. The maximal distance from $\mathbf{v}_2$ is to vertex $\mathbf{v}_8$, and from vertex $\mathbf{v}_4$ to $\mathbf{v}_1$. In both cases the maximal distances are equal and depending on the used norm are

$$\|\mathbf{v}_2 - \mathbf{v}_8\|_1 = (|0 - 1| + |0 - 1| + |1 - 1|) = 2,$$

$$\|\mathbf{v}_2 - \mathbf{v}_8\|_q = (|0 - 1|^q + |0 - 1|^q + |1 - 1|^q)^{\frac{1}{q}} = 2^{\frac{1}{q}}, \quad 1 < q < \infty,$$

$$\|\mathbf{v}_2 - \mathbf{v}_8\|_\infty = \max\{|0 - 1|, |0 - 1|, |1 - 1|\} = 1.$$

The observed situation is analogous to (2.20), therefore by using this triangulation, $\mu_2^{1,\infty}$ bound over three-dimensional simplices is not always better than μ_2^q with one particular q-norm.

After the subdivision of simplices, the situation remains similar: not in all vertices combined bound $\mu_2^{1,\infty}$ is better than μ_2^q type bound with one concrete bound. Therefore for higher dimensional simplices, the real improvement is showed after experimental investigations (see Sect. 2.6).

2.4 Lipschitz Bound Based on Circumscribed Spheres

Calculation of tight Lipschitz bounds is usually computationally expensive. Therefore, it is important to investigate possibilities of bounds tighter than μ_1 (2.10) and μ_2 (2.11), but computationally less expensive than φ (2.8). This section is devoted to the Lipschitz bound over simplices based on circumscribed spheres [100]. The bound is often stronger than usually used trivial bounds and still computationally cheap, especially for low-dimensional problems for which calculation of the radius of the circumscribed sphere is cheap. To find the radius, $n + 2$ determinants of $(n + 1) \times (n + 1)$-dimensionality matrices are calculated.

Proposition 2.2. *Let $f : \mathbb{D} \to \mathbb{R}, \mathbb{D} \subset \mathbb{R}^n$ be a Lipschitz continuous objective function, $\mathbb{I} \subset \mathbb{D}$ be an n-simplex with the set of vertices $\mathbb{V}(\mathbb{I})$, and $R_2(\mathbb{I})$ denotes the radius of the circumscribed n-sphere. Then*

$$\psi^2(\mathbb{I}) = \min_{\mathbf{v} \in \mathbb{V}(\mathbb{I})} f(\mathbf{v}) - L_2 R_2(\mathbb{I}), \tag{2.23}$$

underestimates $f(\mathbf{x}), \forall \mathbf{x} \in \mathbb{I}$.

Proof. An n-simplex $\mathbb{I}$ is covered by n-balls $\mathbb{O}_i, i = 1, \ldots, n + 1$ such that the radius $R_2(\mathbb{I})$ is the same for all n-balls and the centers coincide with the vertices $\mathbf{v}_i$ of the simplex $\mathbb{I}$ (a two-dimensional example is shown in Fig. 2.4): $\forall \mathbf{x} \in \mathbb{I} \exists i$ such that $\mathbf{x} \in \mathbb{O}_i$. Then $\forall \mathbf{x} \in \mathbb{O}_i$

$$f(\mathbf{x}) \geq f(\mathbf{v}_i) - L_2 R_2(\mathbb{I}) \geq \min_{\mathbf{v} \in \mathbb{V}(\mathbb{I})} f(\mathbf{v}) - L_2 R_2(\mathbb{I}) = \psi^2(\mathbb{I}). \tag{2.24}$$

Let us derive a general formula to calculate the radius R_2 of the circumscribed n-sphere. A geometric construction for the radius of the circumscribed 2-sphere (circumscribed circle) is given by Pedoe [104]. The equation of the n-circumsphere of the n-simplex with the vertices $\mathbf{v}_1 = (v_{11}, \ldots, v_{1n}), \ldots, \mathbf{v}_{n+1} = (v_{(n+1)1}, \ldots, v_{(n+1)n})$ expressed in determinant form is

$$\begin{vmatrix} \sum_{i=1}^{n} x_i^2 & x_1 & x_2 & \cdots & x_n & 1 \\ \sum_{i=1}^{n} v_{1i}^2 & v_{11} & v_{12} & \cdots & v_{1n} & 1 \\ \sum_{i=1}^{n} v_{2i}^2 & v_{21} & v_{22} & \cdots & v_{2n} & 1 \\ \vdots & \vdots & \vdots & \ddots & \vdots & \vdots \\ \sum_{i=1}^{n} v_{(n+1)i}^2 & v_{(n+1)1} & v_{(n+1)2} & \cdots & v_{(n+1)n} & 1 \end{vmatrix} = 0. \tag{2.25}$$

Fig. 2.4 Covering of a triangle (two-dimensional simplex) $\mathbb{I}$ by disks (2-balls) $\mathbb{O}_i, i = 1, \ldots, n + 1$ of the same radius $R_2(\mathbb{I})$

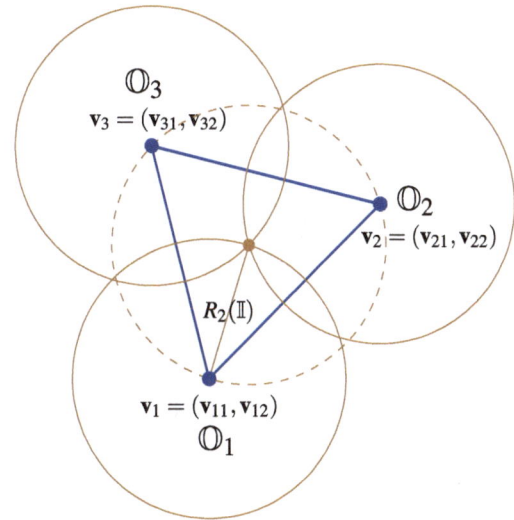

Expanding the determinant by the first row we get

$$\mathbf{a} \left(\sum_{i=1}^{n} x_i^2 \right) + \left(\sum_{i=1}^{n} \mathbf{b}_i x_i \right) + \mathbf{c} = 0, \tag{2.26}$$

where

$$\mathbf{a} = \begin{vmatrix} v_{11} & v_{12} & \cdots & v_{1n} & 1 \\ v_{21} & v_{22} & \cdots & v_{2n} & 1 \\ \vdots & \vdots & \ddots & \vdots & \vdots \\ v_{(n+1)1} & v_{(n+1)2} & \cdots & v_{(n+1)n} & 1 \end{vmatrix},$$

$$\mathbf{b}_1 = - \begin{vmatrix} \sum_{i=1}^{n} v_{1i}^2 & v_{12} & \cdots & v_{1n} & 1 \\ \sum_{i=1}^{n} v_{2i}^2 & v_{22} & \cdots & v_{2n} & 1 \\ \vdots & \vdots & \ddots & \vdots & \vdots \\ \sum_{i=1}^{n} v_{(n+1)i}^2 & v_{(n+1)2} & \cdots & v_{(n+1)n} & 1 \end{vmatrix},$$

$\vdots$

$$
\mathbf{b}_j = (-1)^j
\begin{vmatrix}
\sum_{i=1}^{n} v_{1i}^2 & v_{11} & \cdots & v_{1(j-1)} & v_{1(j+1)} & \cdots & v_{1n} & 1 \\
\sum_{i=1}^{n} v_{1i}^2 & v_{21} & \cdots & v_{2(j-1)} & v_{2(j+1)} & \cdots & v_{2n} & 1 \\
\vdots & \vdots & \ddots & \vdots & \vdots & \ddots & \vdots & \vdots \\
\sum_{i=1}^{n} v_{(n+1)i}^2 & v_{(n+1)1} & \cdots & v_{(n+1)(j-1)} & v_{(n+1)(j+1)} & \cdots & v_{(n+1)n} & 1
\end{vmatrix},
$$

where $j = 2, \ldots, n-1$,

$\vdots$

$$
\mathbf{b}_n = (-1)^n
\begin{vmatrix}
\sum_{i=1}^{n} v_{1i}^2 & v_{11} & \cdots & v_{1n-1} & 1 \\
\sum_{i=1}^{n} v_{1i}^2 & v_{21} & \cdots & v_{2n-1} & 1 \\
\vdots & \vdots & \ddots & \vdots & \vdots \\
\sum_{i=1}^{n} v_{(n+1)i}^2 & v_{(n+1)1} & \cdots & v_{(n+1)n-1} & 1
\end{vmatrix},
$$

$$
\mathbf{c} = (-1)^{n+1}
\begin{vmatrix}
\sum_{i=1}^{n} v_{1i}^2 & v_{11} & \cdots & v_{1n} \\
\sum_{i=1}^{n} v_{1i}^2 & v_{21} & \cdots & v_{2n} \\
\vdots & \vdots & \ddots & \vdots \\
\sum_{i=1}^{n} v_{(n+1)i}^2 & v_{(n+1)1} & \cdots & v_{(n+1)n}
\end{vmatrix}.
$$

Completing the square for (2.26) gives:

$$
\mathbf{a}\left(x_1 + \frac{\mathbf{b}_1}{2\mathbf{a}}\right)^2 + \cdots + \mathbf{a}\left(x_n + \frac{\mathbf{b}_n}{2\mathbf{a}}\right)^2 - \frac{\mathbf{b}_1^2}{4\mathbf{a}} - \cdots - \frac{\mathbf{b}_n^2}{4\mathbf{a}} + \mathbf{c} = 0
$$

which is the n-circumsphere

$$
\left(x_1 + \frac{\mathbf{b}_1}{2\mathbf{a}}\right)^2 + \cdots + \left(x_n + \frac{\mathbf{b}_n}{2\mathbf{a}}\right)^2 = \frac{\mathbf{b}_1^2 + \cdots + \mathbf{b}_n^2 - 4\mathbf{a}\mathbf{c}}{4\mathbf{a}^2}
$$

with the circumcenter

$$\mathbf{c} = (c_1, \ldots, c_n) = \left(-\frac{b_1}{2a}, \ldots, -\frac{b_n}{2a} \right)$$

and the circumradius

$$R_2 = \sqrt{\frac{b_1^2 + \cdots + b_n^2 - 4ac}{4a^2}}.$$

2.5 Tight Lipschitz Bound over Simplices with the 1-norm

In this section, φ^q type bound (2.8) based on the 1-norm (φ^1) is described. In this case, the lower bounding function is the envelope of n-dimensional pyramids and its maximum point is found by solving a system of linear equations [98]. In the case of the Euclidean norm, the lower bounding function is the envelope of n-dimensional cones and its maximum point can be found by solving a system of quadratic and linear equations. Therefore, the bound based on the first norm is less computationally expensive. Let us formulate two propositions, which are used for the evaluation of φ^1.

Proposition 2.3. *Let two n-pyramids $F_{\mathbf{v}_1}(\mathbf{x}) = f(\mathbf{v}_1) - L_\infty \|\mathbf{x} - \mathbf{v}_1\|_1$ and $F_{\mathbf{v}_2}(\mathbf{x}) = f(\mathbf{v}_2) - L_\infty \|\mathbf{x} - \mathbf{v}_2\|_1$ are defined and $f(\mathbf{v}_1) \leq f(\mathbf{v}_2)$. Then the intersection of pyramids is contained in a manifold of dimensionality $n - 1$ defined by*

$$\sum_{i=1}^{n} d(\mathbf{v}_{1i}, \mathbf{v}_{2i}) - \frac{f(\mathbf{v}_2) - f(\mathbf{v}_1)}{L_\infty} = 0, \tag{2.27}$$

where

$$d(\mathbf{v}_{1i}, \mathbf{v}_{2i}) = \begin{cases} |2\mathbf{x}_i - \mathbf{v}_{1i} - \mathbf{v}_{2i}| & \text{when } \mathbf{v}_{1i} \leq \mathbf{x}_i \leq \mathbf{v}_{2i}, \\ 0 & \text{when } \mathbf{v}_{1i} = \mathbf{v}_{2i}, \\ |\mathbf{v}_{1i} - \mathbf{v}_{2i}| & \text{when } \mathbf{x}_i \notin [\mathbf{v}_{1i}, \mathbf{v}_{2i}], \end{cases}$$

and all points $\mathbf{x}$ in the intersection are closer to vertex $\mathbf{v}_1$ than to $\mathbf{v}_2$, i.e.,

$$\|\mathbf{x} - \mathbf{v}_1\|_1 \leq \|\mathbf{x} - \mathbf{v}_2\|_1. \tag{2.28}$$

Proof. From equality $F_{\mathbf{v}_1}(\mathbf{x}) = F_{\mathbf{v}_2}(\mathbf{x})$ we get

$$\sum_{i=1}^{n} (-|\mathbf{x}_i - \mathbf{v}_{1i}| + |\mathbf{x}_i - \mathbf{v}_{2i}|) = \frac{f(\mathbf{v}_2) - f(\mathbf{v}_1)}{L_\infty}. \tag{2.29}$$

For each of the difference $|\mathbf{x}_i - \mathbf{v}_{2i}| - |\mathbf{x}_i - \mathbf{v}_{1i}|$, $i = 1, \ldots, n$ one case from the following possibilities is true:

$$\begin{cases} -2\mathbf{x}_i + \mathbf{v}_{1i} + \mathbf{v}_{2i} & \text{when } \mathbf{v}_{1i} \leq \mathbf{x}_i \leq \mathbf{v}_{2i}, \\ 2\mathbf{x}_i - \mathbf{v}_{1i} - \mathbf{v}_{2i} & \text{when } \mathbf{v}_{2i} \leq \mathbf{x}_i \leq \mathbf{v}_{1i}, \\ 0 & \text{when } \mathbf{v}_{1i} = \mathbf{v}_{2i}, \\ \mathbf{v}_{1i} - \mathbf{v}_{2i} & \text{when } \mathbf{x}_i > \mathbf{v}_{1i}, \mathbf{v}_{2i}, \ \mathbf{v}_{1i} > \mathbf{v}_{2i}, \\ \mathbf{v}_{2i} - \mathbf{v}_{1i} & \text{when } \mathbf{x}_i > \mathbf{v}_{1i}, \mathbf{v}_{2i}, \ \mathbf{v}_{1i} < \mathbf{v}_{2i}. \end{cases}$$

Therefore from (2.29) we get (2.27).

Because $f(\mathbf{v}_1) \leq f(\mathbf{v}_2)$, from the equality $F_{\mathbf{v}_1}(\mathbf{x}) = F_{\mathbf{v}_2}(\mathbf{x})$ we get

$$-\|\mathbf{x} - \mathbf{v}_1\|_1 + \|\mathbf{x} - \mathbf{v}_2\|_1 = \frac{f(\mathbf{v}_2) - f(\mathbf{v}_1)}{L_\infty} \geq 0,$$

therefore (2.28) is true.

Proposition 2.4. *The Piyavskii type Lipschitz bound with the first norm φ^1 can be found solving a system of n linear equations.*

Proof. Let us define the numeration of vertices $\mathbf{v}_i$ so that $f(\mathbf{v}_1) \leq f(\mathbf{v}_2) \leq \cdots \leq f(\mathbf{v}_{n+1})$. Intersection of pyramids $F_{\mathbf{v}_1}(\mathbf{x}) = f(\mathbf{v}_1) - L_\infty \|\mathbf{x} - \mathbf{v}_1\|_1$ and $F_{\mathbf{v}_i}(\mathbf{x}) = f(\mathbf{v}_i) - L_\infty \|\mathbf{x} - \mathbf{v}_i\|_1$, $i = 2, \ldots, n+1$ is $(n-1)$-manifold defined by (2.27). Taking into account (2.28), it is possible to consider only part of the manifold, which is defined by linear equation and constraints. Therefore it is possible to form a system of n linear equations defining intersections $F_{\mathbf{v}_1}(\mathbf{x}) = F_{\mathbf{v}_i}(\mathbf{x})$. If the solution of this system satisfies the constraints (see Example 2.1), then the lower bounding function attains it minimum at the solution point. If the solution of this system does not satisfy the constraints, then the minimum of the lower bounding function is the maximum of the function value at the intersections (see Example 2.2).

Example 2.1. Suppose the objective function is $f(\mathbf{x}) = -\sin(2x_1 + 1) - 2\sin(3x_2 + 2)$ and the feasible region $\mathbb{D} = [0, 1] \times [0, 1]$ is covered by two right-angled isosceles triangles $\mathbb{I}_1 = [\mathbf{v}_1, \mathbf{v}_2, \mathbf{v}_3]$ and $\mathbb{I}_2 = [\mathbf{v}_1, \mathbf{v}_3, \mathbf{v}_4]$ (see Fig. 2.5).

Let us consider the first simplex $\mathbb{I}_1 = [\mathbf{v}_1, \mathbf{v}_2, \mathbf{v}_3]$.

Because $f(\mathbf{v}_1) = f(0, 0) = -2.6601 < f(\mathbf{v}_2) = f(1, 0) = -1.9597 < f(\mathbf{v}_3) = f(1, 1) = 1.7767$, the intersection point is closer to vertex $\mathbf{v}_1$ and it is enough to find the intersection of pyramids $F_{\mathbf{v}_1} = F_{\mathbf{v}_2}$ and $F_{\mathbf{v}_1} = F_{\mathbf{v}_3}$.

The intersection of pyramids $F_{\mathbf{v}_1} = F_{\mathbf{v}_2}$ is:

$$-|\mathbf{x}_1 - 0| + |\mathbf{x}_1 - 1| - |\mathbf{x}_2 - 0| + |\mathbf{x}_2 - 0| = \frac{f(1, 0) - f(0, 0)}{6},$$

however as $0 \leq \mathbf{x}_1 \leq 1$,

$$-\mathbf{x}_1 - \mathbf{x}_1 + 1 = \frac{f(1, 0) - f(0, 0)}{6},$$

Fig. 2.5 Example 2.1: the projection of the intersection lines

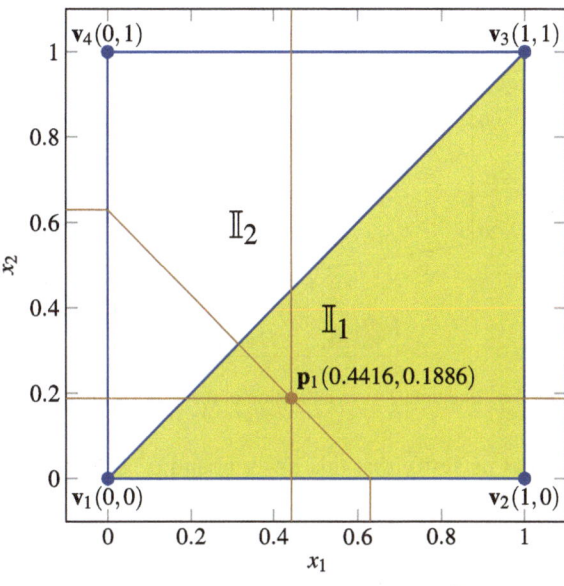

$$\mathbf{x}_1 = 0.4416. \tag{2.30}$$

Analogously intersection of pyramids $F_{\mathbf{v}_1} = F_{\mathbf{v}_3}$ is:

$$-|\mathbf{x}_1 - 0| + |\mathbf{x}_1 - 1| - |\mathbf{x}_2 - 0| + |\mathbf{x}_2 - 1| = \frac{f(1,1) - f(0,0)}{6},$$

$$\begin{cases} \mathbf{x}_1 + \mathbf{x}_2 = -\frac{f(1,1)-f(0,0)}{12} + 1 = 0.6303, & \text{when } 0 \le \mathbf{x}_1 \le 1, \ 0 \le \mathbf{x}_2 \le 1, \\ \mathbf{x}_2 = 0.6303, & \text{when } \mathbf{x}_1 \le 0, \\ \mathbf{x}_1 = 0.6303, & \text{when } \mathbf{x}_2 \le 0. \end{cases} \tag{2.31}$$

From (2.28), (2.30), (2.31) $\Rightarrow$

$$\begin{cases} \mathbf{x}_1 = 0.4416 \\ \mathbf{x}_1 + \mathbf{x}_2 = 0.6303 \\ 0 \le \mathbf{x}_1 \le 0.5 \\ 0 \le \mathbf{x}_2 \le 1.0 \end{cases} \Rightarrow \text{intersection point } \mathbf{p}_1 = (0.4416, 0.1886).$$

Here we give four most significant digits. The lower bound $\varphi^1(\mathbb{I}_1)$ (see Fig. 2.6) is

$$\varphi^1(\mathbb{I}_1) = F_{\mathbf{v}_i}(\mathbf{p}_1) = F_{\mathbf{v}_1}(\mathbf{p}_1) = f(\mathbf{v}_1) - L_\infty \|\mathbf{p}_1 - \mathbf{v}_1\|_1$$

$$= f(0,0) - 6(|0.4416 - 0| + |0.1886 - 0|) = -6.441. \tag{2.32}$$

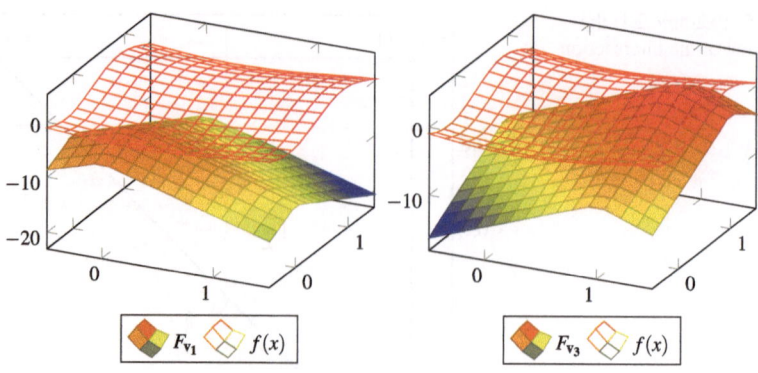

Fig. 2.6 Example 2.1: visualization of the lower bounding functions

Let us verify that this lower bound φ^1 is better than the simpler μ_2^q type bound with the first norm μ_2^1:

$$\mu_2^1(v_1) = f(0,0) - 6 \max_{x \in \mathbb{I}} \|x - v_1\|_1 = f(0,0) - 6 \cdot 2 = -14.66,$$

$$\mu_2^1(v_2) = f(1,0) - 6 \max_{x \in \mathbb{I}} \|x - v_2\|_1 = f(1,0) - 6 \cdot 1 = -7.9597,$$

$$\mu_2^1(v_3) = f(1,1) - 6 \max_{x \in \mathbb{I}} \|x - v_3\|_1 = f(1,1) - 6 \cdot 2 = -10.223,$$

$$\mu_2^1(\mathbb{I}_1) = \max\{\mu_2^1(v_1), \mu_2^1(v_2), \mu_2^1(v_3)\} = -7.9597. \tag{2.33}$$

From (2.32) and (2.33) $\Rightarrow \varphi^1(\mathbb{I}_1) > \mu_2^1(\mathbb{I}_1)$.

Example 2.2. The subdivision of simplices $\mathbb{I}_1$ and $\mathbb{I}_2$ produces four simplices $\mathbb{I}_3, \ldots, \mathbb{I}_6$ (see Fig. 2.7). Let us consider simplex $\mathbb{I}_3 = [v_1, v_2, v_5]$. Since

$$f(v_1) = -2.6601 < f(v_2) = -1.9597 < f(v_5) = -0.2077,$$

the intersection point is closer to vertex v_1 and can be found from intersection of pyramids $F_{v_1} = F_{v_2}$ and $F_{v_1} = F_{v_5}$.

Intersection of pyramids $F_{v_1} = F_{v_2}$ is found in Example 2.1:

$$x_1 = 0.4416. \tag{2.34}$$

Intersection of pyramids $F_{v_1} = F_{v_5}$:

$$-|x_1 - 0| + |x_1 - 0.5| - |x_2 - 0| + |x_2 - 0.5| = \frac{f(0.5, 0.5) - f(0,0)}{6},$$

$$\begin{cases} x_1 + x_2 = -\dfrac{f(0.5, 0.5) - f(0,0)}{12} + \dfrac{1}{2} = 0.2956, & \text{when } 0 \le x_1, x_2 \le 0.5 \\ x_2 = 0.2956, & \text{when } x_1 \le 0, \\ x_1 = 0.2956, & \text{when } x_2 \le 0. \end{cases} \tag{2.35}$$

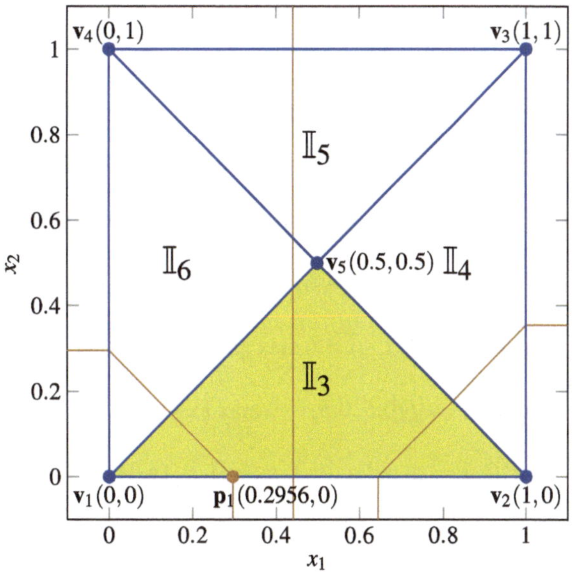

Fig. 2.7 Example 2.2: the
projection of the intersection
lines

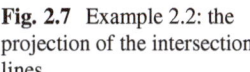

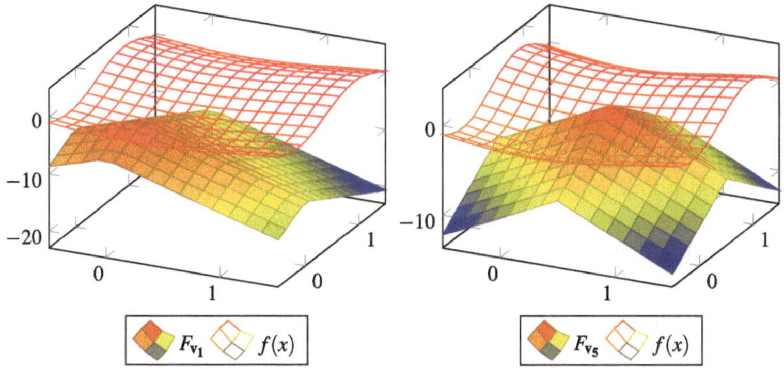

Fig. 2.8 Example 2.2: visualization of the lower bounding functions

From (2.28), (2.34), (2.35) $\Rightarrow$

$$\begin{cases} \mathbf{x}_1 = 0.4416 \\ \mathbf{x}_1 + \mathbf{x}_2 = 0.2956 \\ 0 \le \mathbf{x}_1 \le 0.5 \\ 0 \le \mathbf{x}_2 \le 0.5 \end{cases} \Rightarrow \varnothing \text{ (see Fig. 2.8)}$$

Therefore the lower bound is attained at the intersection line belonging to the lower
bounding function (see Fig. 2.8), i.e.

$$\varphi^1(\mathbb{I}_3) = F_{\mathbf{v}_1}(\mathbf{p}_1),$$

where $\mathbf{p}_1$ is any point belonging to $\mathbf{x}_1 + \mathbf{x}_2 = 0.2956$ and $0 \le \mathbf{x}_1 \le 0.5, 0 \le \mathbf{x}_2 \le 0.5$. Let $\mathbf{p}_1 = (0.2956, 0)$, then

$$\varphi^1(\mathbb{I}_3) = F_{\mathbf{v}_1}(\mathbf{p}_1) = f(\mathbf{v}_1) - L_\infty \|\mathbf{p}_1 - \mathbf{v}_1\|_1$$
$$= f(0,0) - 6(|0.2956 - 0| + |0 - 0|) = -4.4339. \qquad (2.36)$$

Let us verify that this φ^q type bound with the first norm φ^1 is better than μ_2^1:

$$\mu_2^1(\mathbf{v}_1) = f(0,0) - 6 \max_{\mathbf{x} \in \mathbb{I}} \|\mathbf{x} - \mathbf{v}_1\|_1 = f(0,0) - 6 \cdot 1 = -8.6601,$$

$$\mu_2^1(\mathbf{v}_2) = f(1,0) - 6 \max_{\mathbf{x} \in \mathbb{I}} \|\mathbf{x} - \mathbf{v}_2\|_1 = f(1,0) - 6 \cdot 1 = -7.9597,$$

$$\mu_2^1(\mathbf{v}_5) = f(0.5,0.5) - 6 \max_{\mathbf{x} \in \mathbb{I}} \|\mathbf{x} - \mathbf{v}_5\|_1 = f(0.5,0.5) - 6 \cdot 1 = -6.2077,$$

$$\mu_2^1(\mathbb{I}_3) = \max \left\{ \mu_2^1(\mathbf{v}_1), \mu_2^1(\mathbf{v}_2), \mu_2^1(\mathbf{v}_5) \right\} = -6.2077. \qquad (2.37)$$

From (2.36) and (2.37) $\Rightarrow \varphi^1(\mathbb{I}_3) > \mu_2^1(\mathbb{I}_3)$.

2.6 Branch-and-Bound with Simplicial Partitions and Various Lipschitz Bounds

Branch-and-bound algorithms differ by selection strategy of the node to process, branching and bound calculation. A general branch-and-bound algorithm was presented in Algorithm 1. In this section we use simplicial partitioning, combinatorial covering of the rectangular feasible region, best-first selection, subdivision of simplices through the middle of the longest edge, the minimum of the function values at the vertices of simplices as the upper bound for the minimum and various lower bounds μ_2^q, φ^1, ψ^2, as well as aggregates of them with different norms and the corresponding Lipschitz constants. Simplicial branch-and-bound algorithm is shown in Algorithm 4.

Algorithm 4 Simplicial branch-and-bound algorithm with various Lipschitz bounds

---

1: Cover $\mathbb{D}$ by $\mathbb{L} \leftarrow \left\{ \mathbb{L}_j \,|\, \mathbb{D} = \bigcup_{j=1}^{n!} \mathbb{L}_j \right\}$ using combinatorial triangulation
2: $\mathbb{S} \leftarrow \varnothing, UB(\mathbb{D}) \leftarrow \infty$
3: **while** ($\mathbb{L} \ne \varnothing$) **do**
4:     Choose $\mathbb{I} \in \mathbb{L}$ using selection strategy, $\mathbb{L} \leftarrow \mathbb{L} \setminus \{\mathbb{I}\}$
5:     $UB(\mathbb{D}) \leftarrow \min\{UB(\mathbb{D}), \min_{\mathbf{v} \in V(\mathbb{I})} f(\mathbf{v})\}$
6:     $\mathbb{S} \leftarrow \arg\min \left\{ f(\mathbb{S}), \min_{\mathbf{v} \in V(\mathbb{I})} f(\mathbf{v}) \right\}$
7:     Calculate $LB(\mathbb{I})$ using different Lipschitz bounds
8:     **if** ($LB(\mathbb{I}) < UB(\mathbb{D}) - \varepsilon$) **then**
9:         Branch $\mathbb{I}$ into 2 simplices: $\mathbb{I}_1, \mathbb{I}_2$
10:        $\mathbb{I} \leftarrow \mathbb{I} \cup \{\mathbb{I}_1, \mathbb{I}_2\}$
11:    **end if**
12: **end while**

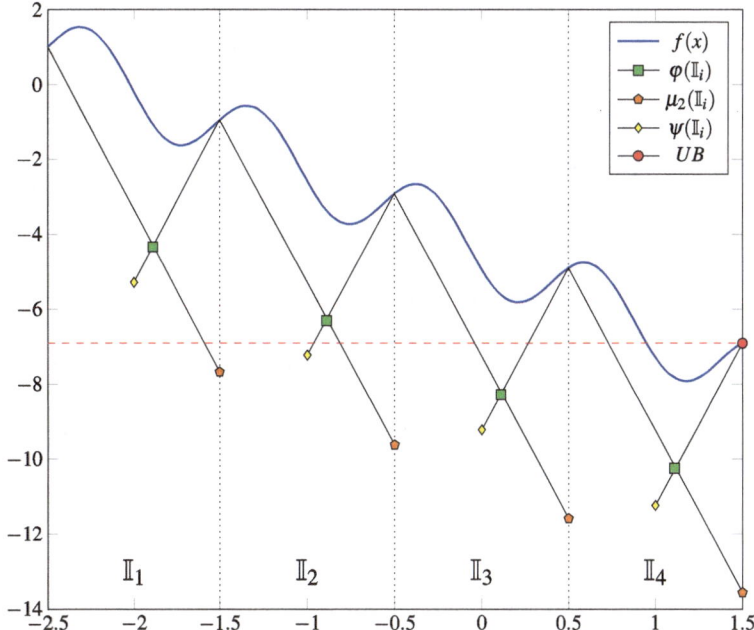

Fig. 2.9 Second step of branch-and-bound algorithm with simplicial partitions ($n = 1$)

The branch-and-bound process is illustrated using an one-dimensional example in Fig. 2.9.

Example 2.3. Let $\mathbb{D} = [-2.5, 1.5]$, $f(x)$ is Lipschitz function (2.9). As the example is one dimensional, initial covering is trivial: the feasible region is the initial simplex. Then it is divided into two through the middle point (the first step) and both are subdivided into two (the second step) again. At this step of the search there are four simplices $\mathbb{I}_i$, $i = 1, \ldots, 4$ (see Fig. 2.9). Three different lower bounds of different type: φ (—■—), μ_2 (—●—) and ψ (—◇—) are shown. We remind that in one-dimensional case all norm are equal; therefore there is no need to specify what norms are used.

The bounds μ_2 over all simplices $\mathbb{I}_i$: $\mu_2(\mathbb{I}_i) < UB$, $i = 1, \ldots, 4$, however using ψ type lower bound we get $\psi(\mathbb{I}_1) > UB$ and using the φ type lower bound we get $\varphi(\mathbb{I}_1) > UB$ and $\varphi(\mathbb{I}_2) > UB$. Therefore, two simplices ($\mathbb{I}_1, \mathbb{I}_2$) in the case of φ type bound and only one simplex $\mathbb{I}_1$ in the case of ψ bound can be discarded from the further search. Only non-discarded simplices will be subdivided further. Therefore we can detect subregions that cannot contain the global optimizer using φ and ψ type bounds faster.

Let us perform and discuss computational experiments with simplicial branch-and-bound algorithm using different Lipschitz bounds. Various test problems ($n \geq 2$) for global optimization from [49, 60, 82] have been used in our experiments. The full description of the used test problems is given in Appendix A. For $(n = 2, 3)$ we use the same precision as used by Hansen and Jaumard [49]. The speed of global optimization has been estimated using the number of function evaluations criterion. Average values for each dimensionality are shown in bold font in following tables.

For two-dimensional test problems no single norm and the corresponding Lipschitz constant are the best for all test functions. However theoretical (see Sect. 2.3) and experimental investigation (see Table 2.3) shows that using the Lipschitz bound based on two extreme (infinite and first) norms $\mu_2^{1,\infty}$ (2.17) gives the best results. On average the number of function evaluations is by 21% smaller than when the bound based on the traditional Euclidean norm μ_2^2 is used.

It was shown in Sect. 2.3 that when the feasible region is a hyper-rectangle and $n \geq 3$, the bound $\mu_2^{1,\infty}$ is not always best. For some optimization problems better results are obtained using other norms. The results of experimental investigation suggest us include in μ_2^2 bound in the aggregate bound:

$$\mu_2^{1,2,\infty}(\mathbb{I}) = \max_{\mathbf{v} \in \mathbb{V}(\mathbb{I})} \{f(\mathbf{v}) - K(\mathbb{V}(\mathbb{I}))\}, \tag{2.38}$$

where

$$K(\mathbb{V}(\mathbb{I})) = \min \left\{ L_\infty \max_{\mathbf{x} \in \mathbb{V}(\mathbb{I})} \|\mathbf{x} - \mathbf{v}\|_1, L_2 \max_{\mathbf{x} \in \mathbb{V}(\mathbb{I})} \|\mathbf{x} - \mathbf{v}\|_2, L_1 \max_{\mathbf{x} \in \mathbb{V}(\mathbb{I})} \|\mathbf{x} - \mathbf{v}\|_\infty \right\}.$$

For $n = 2$ test problems the Euclidean norm is useful only for rectangular feasible regions. The feasible region of only one (No. 10) of the thirteen test problems is rectangle. Only for this test problem $\mu_2^{1,2,\infty}$ improves the result comparing with the result when $\mu_2^{1,\infty}$ was used.

For test problems of higher dimensionality $(n \geq 3)$ the bound $\mu_2^{1,2,\infty}$ improves the results almost for all test problems comparing with $\mu_2^{1,\infty}$ bound. On average the number of function evaluations is about 5% smaller than with $\mu_2^{1,\infty}$ bound and about 52% than in the case the Euclidean norm is used alone.

We observe that using only one single norm μ_2^q for some test problems the solution has not been found after a huge number of function evaluations— 10, 000, 000.

It was mentioned that the Piyavskii type bound (φ) is the tightest (most accurate) Lipschitz bound over some subregion, when the function values at some feasible points and the Lipschitz constant L is known. However calculation of such a bound is a hard optimization problem involving solving a system of quadratic [112] or quadratic and linear [89] equations. Computationally less expensive Piyavskii type bound based on the first norm φ^1 was described in Sect. 2.5.

On the first part of the experiments the bound φ^1 was compared with μ_2^1. From the results presented in Table 2.4, we see that for all test problems better results are achieved using φ^1 than μ_2^1. Depending on the dimensionality of test problems, the number of function evaluations is from 4 to 30% smaller than with a simpler bound.

Table 2.3 The number of function evaluations using μ_2^1, μ_2^2, μ_2^∞, $\mu_2^{1,\infty}$, and $\mu_2^{1,2,\infty}$ bounds

#	μ_2^1	μ_2^2	μ_2^∞	$\mu_2^{1,\infty}$	$\mu_2^{1,2,\infty}$
1	2019	1356	970	970	970
2	321	299	246	216	216
3	10700	8935	10825	7539	7539
4	40	52	72	8	8
5	61	126	153	61	61
6	2864	2046	1953	1751	1751
7	34814	23100	24098	20200	20200
8	528	729	1042	521	521
9	73333	42844	46233	40314	40314
10	2421	3055	4397	2369	2297
11	6783	6986	9426	5654	5654
12	29369	37756	58690	29357	29357
13	44908	37774	22262	22241	22241
mean	**16012**	**12697**	**13874**	**10092**	**10087**
14	4157824	6043338	5693624	4157824	4157740
15	51730	14140	3864	3864	3864
16	>10000000	>10000000	3145728	3145728	3145728
17	1663261	2702422	3721012	1596633	1596633
18	47080	19632	38888	38792	18720
19	25292	50812	60088	24748	24748
20	904186	713398	777482	777478	668510
mean	**>2407053**	**>2791963**	**1920098**	**1392152**	**1373706**
21	>10000000	>10000000	6496801	6496791	6264147
22	256557	225298	721331	239217	175107
23	556488	1048292	5347547	556488	556412
24	1031808	1066128	5353683	1031808	1031808
25	573359	1066128	5359481	573281	573025
26	2036200	1565127	5801659	1852688	1253836
27	572414	496904	3818039	549730	438141
28	408540	489831	1933931	384746	331977
mean	**>1929421**	**>1994714**	**4354059**	**1460594**	**1328057**
29	>10000000	7914387	1917124	1917124	1917092
30	6936368	8284881	>10000000	6932546	6239743
31	6490797	8535473	>10000000	6482327	6099520
mean	**>7809055**	**8244914**	**>7305708**	**5110666**	**4752118**
32	>10000000	6269636	823320	823320	821892
33	1926497	7419819	>10000000	1926497	1919569
mean	**>5963249**	**6844728**	**>5411660**	**1374909**	**1370731**

Previous investigations have shown (see Table 2.3) that there are no single norm and the corresponding Lipschitz constant best for all test problems. Therefore, Piyavskii type bound with the first norm φ^1 is not optimal for all test problems. From the results shown in Table 2.4, it can be seen that φ^1 bound gives worse results than $\mu_2^{1,2,\infty}$ for the test problems, for which the bounds with the first norm

Table 2.4 The number of function evaluations using μ_2^1, φ^1, $\mu_2^{1,2,\infty}$, and $\varphi^1\mu_2^{2,\infty}$ bounds

#	μ_2^1	φ^1	$\mu_2^{1,2,\infty}$	$\varphi^1\mu_2^{2,\infty}$
1	2019	1668	970	967
2	321	215	216	188
3	10700	8264	7539	6653
4	40	28	8	8
5	61	39	61	39
6	2864	2465	1751	1723
7	34814	26792	20200	19171
8	528	381	521	380
9	73333	54630	40314	38860
10	2421	1866	2297	1807
11	6783	5416	5654	4789
12	29369	21160	29357	21153
13	44908	33917	22241	22094
mean	**16012**	**12065**	**10087**	**9064**
14	4157824	3423524	4157740	3423480
15	51730	41212	3864	3863
16	>10000000	>10000000	3145728	3145728
17	1663261	1410165	1596633	1363004
18	47080	34464	18720	16884
19	25292	20876	24748	20487
20	904186	733234	668510	547622
mean	**>2407053**	**>2237639**	**1373706**	**1217295**
21	>10000000	>10000000	6264147	6262233
22	256557	240901	175107	167217
23	556488	536356	556412	536280
24	1031808	988493	1031808	988493
25	573359	544034	573025	543712
26	2036200	1867961	1253836	1176018
27	572414	546240	438141	420417
28	408540	383084	331977	314023
mean	**>1929421**	**>1888384**	**1328057**	**1301049**
29	>10000000	>10000000	1917092	1916941
30	6936368	6730455	6239743	6064924
31	6490797	5641951	6099520	5940304
mean	**>7809055**	**>7682866**	**4752118**	**4640723**
32	>10000000	>10000000	821892	821892
33	1926497	1875911	1919569	1868983
mean	**>5963249**	**>5937956**	**1370731**	**1345438**

are not efficient. Therefore, in order to reduce the number of function evaluations, it is appropriate to replace μ_2^1 bound with more tight bound with the same norm (φ^1) in an aggregate. In such a case the aggregate bound may be composed of $\mu_2^{2,\infty}$ and φ^1 [99]:

$$\varphi^1 \mu_2^{2,\infty}(\mathbb{I}) = \max\left\{\varphi^1(\mathbb{I}), \mu_2^{2,\infty}(\mathbb{I})\right\} \tag{2.39}$$

$$= \max\left\{\min_{\mathbf{x}\in\mathbb{I}}\left(\max_{\mathbf{v}\in\mathbb{V}(\mathbb{I})}\{f(\mathbf{v}) - L_\infty\|\mathbf{x}-\mathbf{v}\|_1\}\right), \max_{\mathbf{v}\in\mathbb{V}(\mathbb{I})}\{f(\mathbf{v}) - K'(\mathbb{V}(\mathbb{I}))\}\right\},$$

where

$$K'(\mathbb{V}(\mathbb{I})) = \min\left\{L_1 \max_{\mathbf{x}\in\mathbb{I}}\|\mathbf{x}-\mathbf{v}\|_\infty, L_2 \max_{\mathbf{x}\in\mathbb{I}}\|\mathbf{x}-\mathbf{v}\|_2\right\}.$$

On average the number of function evaluations with this aggregate bound are 9% smaller compared to that of $\mu_2^{1,2,\infty}$, and by 25% smaller compared to the results with φ^1.

The experimental results of simplicial branch-and-bound algorithm with the bound ψ^2 (2.23) and with the bound μ_2^2 (2.10) based on the same Euclidean norm are shown in Table 2.5. The ratio $r_{\psi^2<\mu_2^2}$ shows how often the bound ψ^2 is better (more tight) than the bound μ_2^2. For most of the test problems the number of function evaluations is smaller when the bound ψ^2 is used and on the average it is by 47% smaller for $(n = 2, 3)$ test problems and by 26% smaller for $(n \geq 4)$ test problems than when the bound μ_2^2 is used. However, the bound ψ^2 is not always better comparing with μ_2^2 (the ratio $r_{\psi^2<\mu_2^2}$ is not always equal to 1), therefore for computationally expensive functions it might be worthwhile include in the bound ψ^2 in the aggregate bound $\varphi^1\mu_2^{2,\infty}$ (2.39) resulting in improved aggregate bound $\varphi^1\psi^2\mu_2^{2,\infty}$ [100]:

$$\varphi^1\psi^2\mu_2^{2,\infty}(\mathbb{I}) = \max\left\{\varphi^1\mu_2^{2,\infty}(\mathbb{I}), \psi^2(\mathbb{I})\right\}. \tag{2.40}$$

The number of function evaluations are on average 20% smaller when the improved aggregate bound $\varphi^1\psi^2\mu_2^{2,\infty}$ is used than when the aggregate bound $\varphi^1\mu_2^{2,\infty}$ is used (Table 2.5).

The value of the objective function at the same point is required when computing bounds over neighbor simplices. If this value is evaluated for the parent simplex, it is not necessary to evaluate it again. However, in multidimensional case the same points may be encountered when subdividing different simplices. One possibility is to maintain a list of already evaluated points and before evaluating a point to check it has already been evaluated. Vertex verification reduces the number of function evaluations essentially, and also increases computational time; therefore it is more suitable in the case of computationally expensive functions. The number of function evaluations when improved aggregate bound is used and vertices are verified $\widehat{\varphi^1\psi^2\mu_2^{2,\infty}}$ are shown in Table 2.5 too.

Let us compare the simplicial branch-and-bound algorithm with aggregate bound (2.40) and vertex verifications $\widehat{\varphi^1\psi^2\mu_2^{2,\infty}}$ to other well-known algorithms

Table 2.5 The number of function evaluations using μ_2^2 and ψ^2 type bounds, aggregate $\varphi^1\mu_2^{2,\infty}$ and $\varphi^1\psi^2\mu_2^{2,\infty}$ bounds and $\widehat{\varphi^1\psi^2\mu_2^{2,\infty}}$

#	μ_2^2	ψ^2	$\varphi^1\mu_2^{2,\infty}$	$\varphi^1\psi^2\mu_2^{2,\infty}$	$\widehat{\varphi^1\psi^2\mu_2^{2,\infty}}$	r_{ψ^2/μ_2^2}
1	1356	856	967	716	412	0.93
2	299	286	188	185	122	0.39
3	8935	5048	6653	4843	2551	1.00
4	52	128	8	8	5	0.02
5	126	155	39	39	36	0.06
6	2046	1284	1723	1241	691	0.99
7	23100	12547	19171	12195	6240	0.99
8	729	479	380	341	212	0.96
9	42844	22038	38860	21724	11049	1.00
10	3055	1734	1807	1495	830	0.98
11	6986	3915	4789	3598	1931	0.00
12	37756	19211	21153	16938	8926	1.00
13	37774	20366	22094	18737	9855	1.00
mean	**12697**	**6773**	**9064**	**6312**	**3297**	**0.72**
14	6043338	2642392	3423480	2355490	495711	0.99
15	14140	11928	3863	3784	1030	0.64
16	>10000000	6325269	3145728	3145728	536761	1.00
17	2702422	1131286	1363004	1020782	229049	0.95
18	19632	14368	16884	12032	3091	0.75
19	50812	20776	20487	17105	4684	0.91
20	713398	281998	547622	262311	55605	0.98
mean	**>2791963**	**1489717**	**1217295**	**973890**	**189419**	**0.89**
21	>10000000	7359930	6262233	5349848	540697	0.64
22	225298	182447	167217	120144	13707	0.59
23	1048292	635716	536280	450084	44421	0.67
24	1066128	635832	988493	563835	68502	0.67
25	1066128	635922	543712	465955	44991	0.67
26	1565127	965474	1176018	749518	52078	0.65
27	496904	426493	420417	333568	5769	0.64
28	489831	918077	314023	271614	37981	0.24
mean	**>1994714**	**1469986**	**1301049**	**1038071**	**101018**	**0.60**
29	7914387	5826460	1916941	1633849	84406	0.57
30	8284881	8079412	6064924	4590448	162989	0.52
31	8535473	11291062	5940304	5192437	256963	0.47
mean	**8244914**	**8398978**	**4640723**	**3805578**	**168119**	**0.52**
32	6269636	1623674	821892	524940	9840	0.57
33	7419819	6818423	1868983	1685793	25398	0.50
mean	**6844728**	**4221049**	**1345438**	**1105367**	**17619**	**0.53**

for Lipschitz optimization, described by Hansen and Jaumard [49]. Two classes of algorithms are considered:

1. Algorithms using a single lower-bounding function, i.e., variants of Piyavskii algorithm [111, 112]:

   - Mladineo (Mla) [89],
   - Jaumard, Herrmann, and Ribault (JHR) [61],
   - Wood (Wood) [141, 142].

2. Branch-and-bound algorithms:

   - Simplicial branch-and-bound algorithm with aggregate Lipschitz bound $\overline{\varphi^1 \psi^2 \mu_2^{2,\infty}}$,
   - Galperin (Gal85, Gal88) [37, 38],
   - Pinter (Pint) [107],
   - Meewella and Mayne (MM) [86],
   - Gourdin, Hansen, and Joumard (GHJ) [45].

The comparison of the algorithms is based on the number of function evaluation criterion. The number of function evaluations are presented in Table 2.6.

It is mentioned in [49] that the results for all algorithms may be obtained only when the required precision (ε) is not too restrictive. Even so, some problems cannot be solved by some of the algorithms in reasonable computational time and/or memory size. In the experiments we apply the precision used in [49]. The number of function evaluations are smallest, when the algorithms of Mladineo and of Jaumard, Herrmann, and Ribault are used. However, these algorithms belong to the first class of algorithms and require a longer computational time. The branch-and-bound algorithms require a larger number of function evaluations, but much shorter computational time. The best numbers for branch-and-bound algorithms are shown in bold. The performance of the simplicial branch-and-bound algorithm with aggregate bound $\overline{\varphi^1 \psi^2 \mu_2^{2,\infty}}$ is similar to that of the best branch-and-bound algorithm (GHJ) and often it is even better.

2.7 Parallel Branch-and-Bound with Simplicial Partitions

Global optimization algorithms are computationally intensive and therefore parallel computing is important [16, 19, 28, 88, 94]. For some parallel Lipschitz global optimization methods, conditions, which guarantee considerable speedup with respect to the sequential version of the algorithm, are established [122, 133, 134].

In this section two parallel algorithms based on the branch-and-bound technique are described [103]. OpenMP [13, 14] and MPI [47, 131] were used to implement parallel versions of the algorithm. The parallel algorithms belong to the second type of parallelism according to [40]. The parallel OpenMP versions are implemented as

Table 2.6 Comparison with well-known Lipschitz optimization algorithms

	Iterative algorithms			Branch-and-bound algorithms					
#	Mla	JHR	Wood	$\varphi^1 \psi^2 \mu_2^{2,\infty}$	Gal85	Gal88	Pint	MM	GHJ
1	320	323	5528	**412**	3553	1713	3807	1749	643
2	80	80	2861	**122**	1036	577	1762	744	167
3	2066	2066	70955	**2551**	24214	16089	28417	10839	3531
4	6	6	157	**5**	106	73	1527	94	45
5	41	41	209	**36**	430	217	907	424	73
6	548	548	14740	**691**	7729	2929	7772	2684	969
7	–	5088	183759	**6240**	43123	34705	62917	22799	7969
8	177	177	1403	**212**	2113	1289	2272	964	301
9	–	8838	309763	**11049**	57814	49873	88932	53549	13953
10	673	673	18613	**830**	8508	5628	9022	3814	1123
11	1613	1613	53348	**1931**	18235	12737	20312	9224	2677
12	–	8414	470200	**8926**	63088	56177	105572	45389	12643
13	–	9617	–	**9855**	65536	59049	109227	35949	15695
14	>460	>41700	–	495711	5383113	3886897	–	–	**215061**
15	>290	9363	–	**1030**	635909	347075	–	–	24249
16	>290	>12000	–	**536761**	15620627	–	–	–	1297205
17	>280	>14400	–	**229049**	12481708	–	–	–	268279
18	>690	1309	–	**3091**	46411	23765	–	–	3219
19	446	445	–	**4684**	35463	18669	–	–	7177

synchronous single pool (SSP) (one list of candidate simplices is maintained) and the MPI versions as asynchronous multiple pool (AMP) (different processors use separate lists). The synchronous algorithms are executed in phases.

A sequential branch-and-bound algorithm with simplicial partitions and various Lipschitz bounds was described in Sect. 2.6.

Two OpenMP versions were implemented: with a vertex point verification in the global vertex set of all previously evaluated function values at the midpoints of the longest edge and without it. If the function value at the new vertex point has not been evaluated before, the function is evaluated. In the other case the previously evaluated function value is used to calculate bounds. Therefore it is possible to reduce the number of function evaluations avoiding several evaluations at the same point.

The OpenMP version of the parallel branch-and-bound algorithm with vertex point verification is shown in Algorithm 5. Data parallelism is used. The feasible region $\mathbb{D}$ is subsequently divided into a set of simplices $\mathbb{L} = \{\mathbb{L}_j\}$. In C/C++, OpenMP directives are specified by using the #pragma mechanism. The #pragma directives offer a way for each compiler to offer machine- and operating system-specific features while retaining overall compatibility with the C and C++ languages. Directive "for" specifies that the iterations of the loop immediately following it must be executed in parallel by different threads. "schedule(static)" describes that iterations of the loop are evenly (if possible) divided and then

---

Algorithm 5 OpenMP version of parallel branch-and-bound algorithm

---

1: Cover $\mathbb{D}$ by $\mathbb{L} = \{\mathbb{L}_j | \mathbb{D} \subseteq \bigcup_{j=1}^{n!} \mathbb{L}_j \}$ using face-to-face vertex triangulation.
2: $\mathbb{S} \leftarrow \varnothing$, $UB(\mathbb{D}) \leftarrow \infty$, $\mathbb{V}(\mathbb{D}) = \{\mathbf{v}_j : j = 1, \ldots, 2^n\}$
3: **while** ($\mathbb{L} \neq \varnothing$) **do**
4:     **#pragma omp parallel private** (LB)
5:     **#pragma omp for schedule(static)**
6:     **for** ($j = 1$; $j <= |\mathbb{L}|$; $j++$) **do**
7:         $\mathbb{I} \leftarrow \mathbb{L}$, $\mathbb{L} \leftarrow \varnothing$
8:         **#pragma omp critical** ($UB(\mathbb{D})$)
9:         $UB(\mathbb{D}) \leftarrow \min \{UB(\mathbb{D}), \min_{\mathbf{v} \in \mathbb{V}(\mathbb{I}_j)} \{f(\mathbf{v})\}\}$
10:        **#pragma omp critical** ($\mathbb{S}$)
11:        $\mathbb{S} \leftarrow \arg \min \{f(\mathbb{S}), \min_{\mathbf{v} \in \mathbb{V}(\mathbb{I}_j)} f(\mathbf{v})\}$
12:        Calculate $LB(\mathbb{I}_j)$ using different Lipschitz bounds
13:        **if** ($LB(\mathbb{I}_j) < UB(\mathbb{D}) - \varepsilon$) **then**
14:            Branch $\mathbb{I}_j$ into 2 simplices: $\mathbb{I}_1, \mathbb{I}_2$
15:            **if** $\mathbf{v}_j \notin \mathbb{V}(\mathbb{D})$ **then**
16:                $\mathbb{V}(\mathbb{D}) \leftarrow \mathbb{V}(\mathbb{D}) \cup \{\mathbf{v}_j\}$
17:            **end if**
18:            **#pragma omp critical** ($\mathbb{L}$)
19:            $\mathbb{L} \leftarrow \mathbb{L} \cup \{\mathbb{I}_1, \mathbb{I}_2\}$
20:        **end if**
21:    **end for**
22: **end while**

---

statically assigned to threads. The directive "critical" specifies a region of code that must be executed by only one thread at a time.

Each simplex is evaluated by checking if it can contain an optimal solution. For this purpose, the lower bound $LB(\mathbb{I}_j)$ for the objective function f is evaluated over each simplex and compared with the upper bound $UB(\mathbb{D})$ for the minimum value over the feasible region. If $UB(\mathbb{D}) - LB(\mathbb{I}_j) > \varepsilon$, then the simplex $\mathbb{I}_j$ cannot contain a function value better (smaller) than one already found by more than the given precision ε, and therefore it is rejected. Otherwise it is inserted into the set of unexplored simplices $\mathbb{L}$. The algorithm terminates when there are no more potential simplices to investigate.

Two MPI versions with static load balancing were implemented using a parallel branch-and-bound template [2]: with interchange of the best currently found values among processors and without. When the template is used, only algorithm-specific rules should be described by the user and the standard parts are implemented in the template. The MPI library is used for the underlying communications.

Static load balancing is used: tasks are initially distributed evenly (if possible) among p parallel processes and then the processes work independently and do not exchange any later generated tasks. Each parallel process runs the same algorithm, which is shown in Algorithm 6. The algorithm is very similar to the sequential Algorithm 4. The differences are:

Algorithm 6 MPI version of parallel branch-and-bound algorithm

---

1: Cover $\mathbb{D}$ by $\mathbb{L} = \{\mathbb{L}_j | \mathbb{D} \subseteq \bigcup_{j=1}^{n!} \mathbb{L}_j\}$ using face-to-face vertex triangulation.
2: Perform sequential Algorithm 4 while the cardinality $|\mathbb{L}| < 4p$
3: Evenly (if possible) divide $\mathbb{L}$ among the p processes $\mathbb{L} = \bigcup_{i=1}^{p} \mathbb{L}^i$, $|\mathbb{L}^p| \approx |\mathbb{L}|/p$
4: $\mathbb{S}^p \leftarrow \varnothing, UB(\mathbb{L}^p) \leftarrow \infty$.
5: **while** (all $\mathbb{L}^p$ are not empty: $\mathbb{L}^p \neq \varnothing$) **do**
6:      Choose $\mathbb{I}^p \in \mathbb{L}^p$ using selection rule, $\mathbb{L}^p \leftarrow \mathbb{L}^p \backslash \{\mathbb{I}^p\}$
7:      $UB(\mathbb{L}^p) \leftarrow \min\{UB(\mathbb{L}^p), \min_{\mathbf{v}\in V(\mathbb{I}^p)}\{f(\mathbf{v})\}\}$
8:      Share $UB(\mathbb{L}^p)$ among processors
9:      $\mathbb{S}^p \leftarrow \arg\min\{f(\mathbb{S}^p), \min_{\mathbf{v}\in V(\mathbb{I}^p)} f(\mathbf{v})\}$
10:     Calculate $LB(\mathbb{I}^p)$ using different Lipschitz bounds
11:     **if** $(LB(\mathbb{I}^p) < UB(\mathbb{L}^p) - \varepsilon)$ **then**
12:         Branch $\mathbb{I}^p$ into 2 simplices: $\mathbb{I}_1^p, \mathbb{I}_2^p$
13:         $\mathbb{L}^p \leftarrow \mathbb{L}^p \cup \{\mathbb{I}_1^p, \mathbb{I}_2^p\}$
14:     **end if**
15: **end while**
16: Collect results $\mathbb{S}^p$

- At initial step, just one process performs sequential branch-and-bound algorithm until the number of unexamined simplices $|\mathbb{L}|$ becomes at least four times bigger than the number of MPI processes. Then these simplices are randomly distributed among all the processes. In this case, the sequential part takes a little longer, but decreases influence of static load balancing.
- The best currently found value of the objective function $UB(\mathbb{L}^p)$ is local, but the processes interchange it in one version of the algorithm whenever a better function value is found.
- The results are collected after the branch-and-bound process finished. The best found solution is a numerical approximation of the global solution.

Computational experiments of both OpenMP algorithm versions were performed on the parallel machine Ness (http://www.epcc.ed.ac.uk/facilities/ness/) at Edinburgh Parallel Computing Center. Ness has a shared-memory architecture which allows users the option to run large threaded jobs (e.g., OpenMP) as well as message-passing jobs. The system has two back-end X4600 symmetric multiprocessor (SMP) nodes, both containing 16 processor-cores (2.6 GHz AMD Opteron (AMD64e)) with 2GB of memory per core. Up to 16 processor-cores have been used in the experiments.

The performance of both MPI algorithm versions were tested on a cluster of personal computers (Vilkas (http://vilkas.vgtu.lt/) cluster at Vilnius Gediminas Technical University). It consists of 14 Intel Core 2 Quad Q6600 (2.4GHz, 4 cores) and 9 Intel Core i7-860 (2.8GHz, 4 cores, 8 threads) computers interconnected via Gigabit Ethernet Switch. Up to 16 Intel Core i7-860 cores have been used in the experiments.

The parallel versions of the algorithms were evaluated using a commonly used criterion of parallel algorithms: speedup $s_p = t_1/t_p$ and efficiency of parallelization $e_p = s_p/p$, where t_p is time used by the algorithm implemented on p processes.

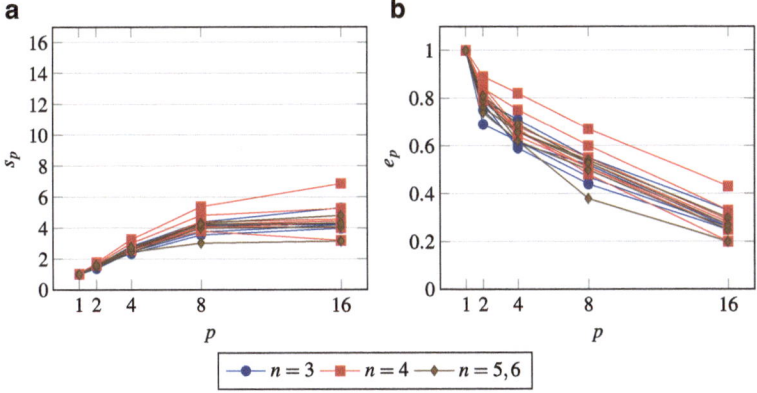

Fig. 2.10   Parallel OpenMP version with vertex verification: (**a**) speedup, (**b**) efficiency

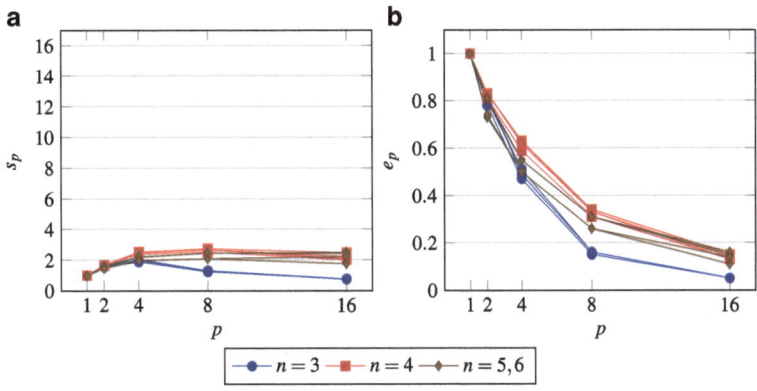

Fig. 2.11   Parallel OpenMP version without vertex verification: (**a**) speedup, (**b**) efficiency

For all parallel versions the aggregate lower bound $\varphi^1 \mu_2^{2,\infty}$ (2.39) with the best first selection strategy was used. Optimization of some problems takes less than a second on a single processor. It is not worthwhile to parallelize the solution of such problems. Therefore in the parallel experiments only more difficult test problems (with a solution time on a single processor of more than 1 s) are used. For test problems $\# = 32, 33$ higher accuracy than in sequential investigations was used.

The diagrams of criteria of parallelization: for various numbers of processes and various dimensionality (n) of test problems are shown in Figs. 2.10–2.13. The diagrams show that the efficiency of parallelization for all parallel versions of algorithm is better for higher dimensional test problems, and much better efficiency is achieved using the MPI versions. The possible reason for this is that significantly less time is spent on communications. The efficiency of OpenMP version with vertex verification is higher than without it.

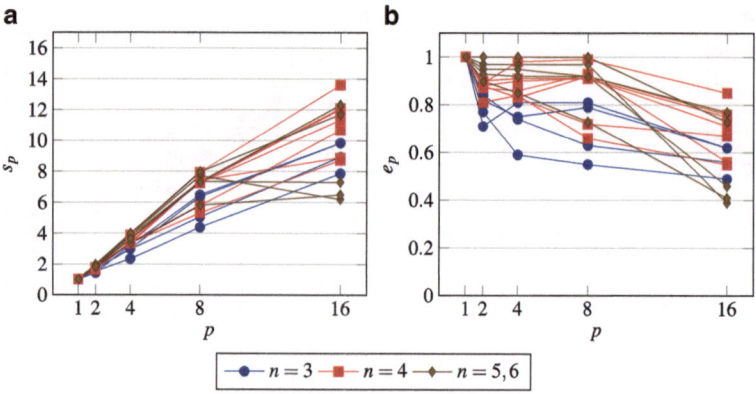

Fig. 2.12 Parallel MPI version without interchange of the best currently found function value: (**a**) speedup, (**b**) efficiency

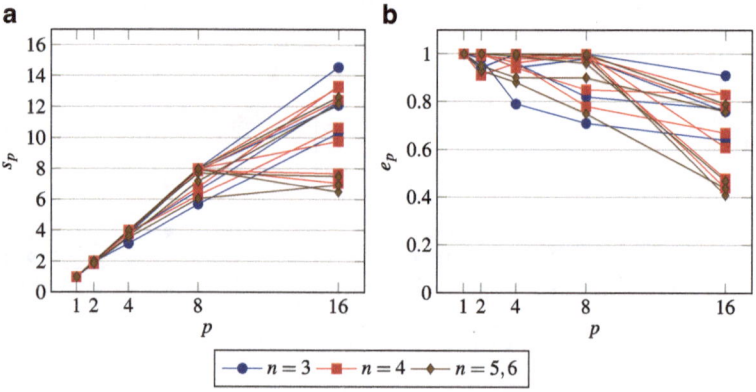

Fig. 2.13 Parallel MPI version with interchange of the best currently found function value: (**a**) speedup, (**b**) efficiency

For both OpenMP parallel versions the number of function evaluations are similar for all numbers of processors and are almost identical to sequential results presented in Table 2.5. For both MPI parallel versions the number of function evaluations using multiple processes are shown in Table 2.7. For most test problems the number of function evaluations using the MPI version with interchange of the best currently found function values (MPI-i) is very similar compared with the MPI version without interchange. For some test problems the number of function evaluations using MPI with interchange is up to 5% smaller than that using MPI without interchange. However, the optimization time using MPI with interchange is almost always longer than for MPI without interchange and is on average about 10% longer (Table 2.8).

Table 2.7 The number of function evaluations using MPI versions on multiple processes

	MPI				MPI-i			
#	$2p.$	$4p.$	$8p.$	$16p.$	$2p.$	$4p.$	$8p.$	$16p.$
14	3423480	3423480	3423480	3423480	3423480	3423480	3423480	3423480
16	3145728	3145728	3145728	3145728	3145728	3145728	3145728	3145728
17	1363119	1363700	1427885	1425216	1363025	1363002	1363010	1363019
20	547622	547622	547981	548079	547622	547622	547622	547622
21	6262233	6262233	6262233	6264548	6262211	6262191	6262224	6262191
22	167217	167217	167217	167217	167217	167217	167217	167217
23	536280	536282	536300	541431	536148	536030	536068	535241
24	988493	988493	988493	988493	988493	988493	988493	988493
25	543712	543716	543717	628371	543451	542736	542984	539629
26	889508	889510	889899	891087	889496	889500	889492	889467
28	314025	314195	316543	318247	313579	313702	313478	313189
29	1916941	1916941	1916941	1916941	1916941	1916941	1916941	1916941
30	6064924	6064924	6064924	6064924	6064924	6064924	6064924	6064924
31	5940304	5940304	5940304	5940304	5940034	5939938	5939645	5938389
32	1488939	1488939	1488939	1488939	1488939	1488939	1488939	1488939
33	11948747	11948747	11948747	11948747	11948747	11948747	11948747	11948747

Table 2.8 Execution time (s) using MPI versions on multiple processes

	MPI				MPI-i			
#	$2p.$	$4p.$	$8p.$	$16p.$	$2p.$	$4p.$	$8p.$	$16p.$
14	5.04	2.63	1.54	0.82	5.77	3.27	1.93	1.09
16	4.92	2.12	1.06	0.63	6.52	2.86	1.43	0.94
17	1.87	1.00	0.47	0.31	2.39	1.31	0.62	0.40
20	0.67	0.42	0.23	0.12	0.86	0.56	0.30	0.17
21	20.37	10.32	4.85	3.28	23.68	12.52	5.65	3.43
22	0.44	0.23	0.11	0.09	0.51	0.26	0.12	0.10
23	1.49	0.75	0.36	0.24	1.78	0.87	0.43	0.26
24	2.96	1.48	0.70	0.45	4.04	1.80	0.83	0.53
25	1.55	0.76	0.37	0.40	1.78	0.88	0.44	0.28
26	2.52	1.35	0.75	0.38	2.90	1.53	0.88	0.48
28	0.99	0.47	0.29	0.17	1.13	0.54	0.34	0.21
29	26.18	13.06	6.66	8.16	27.02	13.55	6.79	8.44
30	80.69	40.57	20.86	21.36	84.67	42.14	21.80	21.87
31	87.29	46.13	26.99	23.40	90.24	47.91	28.04	25.02
32	126.00	63.17	31.58	20.00	125.25	63.24	31.62	21.52
33	1015.47	531.86	267.25	157.45	1040.17	529.17	263.96	157.17

2.8   Experimental Comparison of Selection Strategies

The selection strategies have been introduced in Sect. 1.1. Speed and memory requirements of branch-and-bound algorithms depend on the selection strategy which candidate node to process next [102]. The goal of this section is to experimentally investigate this influence on the performance of sequential and parallel branch-and-bound algorithms. The experiments have been performed solving the same test problems for global optimization as in the previous sections which are described in Appendix A. Branch-and-bound algorithm with simplicial partitions and combination of Lipschitz bounds has been investigated. However, similar results may be expected for other branch-and-bound algorithms.

Parallel experiments have been performed on the computer cluster Ness at Edinburgh Parallel Computing Center (EPCC). It consists of a cluster of two SMP boxes that form the back-end: 2.6 GHz AMD Opteron (AMD64e) processors with 2 GB of memory (32 processors divided into two 16-processors boxes), and a two-processor front-end. The computer cluster runs Linux operating system (Scientific Linux) and Sun Grid Engine.

Let us start the investigation from the sequential branch-and-bound algorithm with simplicial partitions and improved aggregate Lipschitz bound $\varphi^1 \psi^2 \mu_2^{2,\infty}$ (2.40). The number of function evaluations (fe) and execution time ($t(s)$) using different selection strategies are shown in Table 2.9.

For 2-dimensional test problems the *depth first* selection strategy is the least efficient. For test problems with higher dimensionality $n \geq 3$ the average number of function evaluations are very similar for all selection strategies and the differences are insignificant. For test problems of all dimensionalities the smallest execution time is achieved when *depth first* and *breath first* selection strategies are used, despite the fact that sometimes the number of function evaluations is higher. The reason is that the time of insertion and deletion of elements to/from such a type of structure does not depend on the number of elements in the list. *Best first* and *statistical* selection strategies require prioritized list of candidates, and with the heap structure the insertion is time consuming when the number of elements in the list is large.

The speed of locating the global solution is measured by using the ratio

$$r(f^*) = \frac{fe(f^*)}{fe}, \tag{2.41}$$

where $fe(f^*)$ is the number of function evaluations after which the best global solution f^* is found and fe is the number of function evaluations during the branch-and-bound process. The value of the ratio is between zero and one and shows how fast the global solution f^* is found during the optimization process. The ratio $r(f^*)$ for all test problems is shown in Fig. 2.14. The average ratios $\overline{r(f^*)}$ for test problems of different dimensionalities are shown in Table 2.10. For almost all test problems the smallest ratio is achieved when the *statistical* selection strategy is used and average ratios are more than two times smaller for this strategy comparing with other selection strategies. For test problems of dimensionalities $n = 2$ and $n = 3$ the

Table 2.9 The number of function evaluations and execution time for different selection strategies

#	Best first fe	t(s)	Depth first fe	t(s)	Breadth first fe	t(s)	Statistical fe	t(s)
1	716	0.002	4513	0.013	720	0.002	707	0.002
2	185	0.001	199	0.000	185	0.001	184	0.001
3	4843	0.014	5312	0.015	4844	0.014	4843	0.015
4	8	0.000	8	0.000	8	0.000	8	0.000
5	39	0.000	39	0.000	53	0.001	39	0.000
6	1241	0.003	5921	0.016	1241	0.003	1241	0.003
7	12195	0.037	12296	0.034	12195	0.034	12202	0.041
8	341	0.001	462	0.001	342	0.001	340	0.001
9	21724	0.068	21760	0.062	21724	0.058	21724	0.066
10	1495	0.005	1606	0.005	1511	0.004	1493	0.005
11	3598	0.010	5195	0.015	3601	0.009	3979	0.012
12	16938	0.052	17001	0.046	16939	0.046	16937	0.053
13	18737	0.062	18781	0.056	18747	0.055	18766	0.066
mean	**6312**	**0.020**	**7161**	**0.020**	**6316**	**0.018**	**6343**	**0.020**
14	2355490	19.73	2355490	16.90	2355490	17.11	2355490	19.02
15	3784	0.03	6442	0.05	4193	0.03	3797	0.03
16	3145728	27.53	3145728	23.10	3145728	23.15	3145728	27.22
17	1020782	8.58	1059124	7.79	1022743	7.62	1020772	8.50
18	12032	0.09	15025	0.11	12321	0.09	12160	0.10
19	17105	0.14	108820	0.78	17105	0.13	17105	0.13
20	262311	2.09	262340	1.89	262351	1.89	262308	2.07
mean	**973890**	**8.31**	**993281**	**7.23**	**974276**	**7.15**	**973909**	**8.15**
21	5349848	190.17	5350803	185.14	5349944	185.65	5349782	191.10
22	120144	4.13	120144	4.12	120144	4.00	120144	4.15
23	450084	15.73	443780	15.17	443219	15.02	443588	15.74
24	563835	20.02	563836	19.34	563835	19.20	563835	19.96
25	465955	15.81	445906	15.23	443433	15.27	445309	15.91
26	749518	26.27	750735	25.43	750092	25.35	749851	26.26
27	333568	11.79	333471	11.30	333512	11.45	333470	11.63
28	271614	9.38	275474	9.46	276474	9.47	269485	9.35
mean	**1038071**	**36.66**	**1035519**	**35.65**	**1035082**	**35.68**	**1034433**	**36.76**
29	1633849	307.36	1633969	316.83	1633837	310.06	1633837	305.74
30	4590448	864.75	4590448	849.56	4590448	856.56	4590448	879.39
31	5192437	976.94	5202842	976.77	5222690	970.65	5189886	960.17
mean	**3805578**	**716.35**	**3809086**	**714.39**	**3815658**	**712.42**	**3804724**	**715.10**
32	524940	805.94	524940	794.75	524940	779.13	524940	829.04
33	1685793	2451.97	1685793	2462.63	1685793	2464.22	1685793	2578.57
mean	**1105367**	**1324.75**	**1105367**	**1323.92**	**1105367**	**1318.59**	**1105367**	**1374.24**

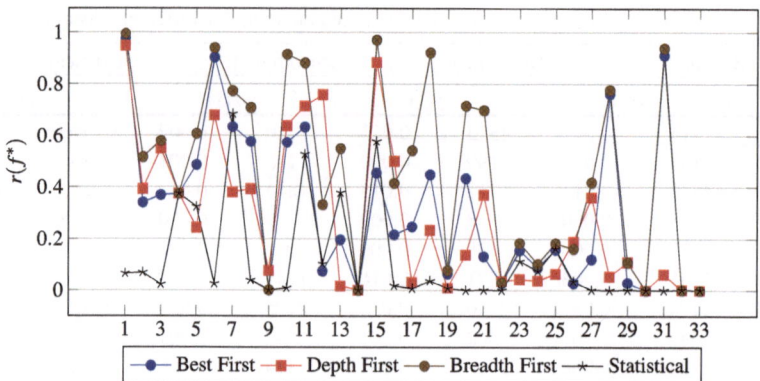

Fig. 2.14 The ratio $r(f^*)$ for the algorithms with different selection strategies

Table 2.10 Average ratio $r(f^*)$ for the algorithms with different selection strategies

n	Best first	Depth first	Breadth first	Statistical
2	0.47	0.47	0.63	0.20
3	0.27	0.26	0.52	0.09
4	0.18	0.14	0.32	0.05
5–6	0.19	0.03	0.21	0.00

ratios are very similar for the *best first* and *depth first* search strategies, but for $n \geq 4$ better results (smaller ratio) are achieved when the *depth first* selection strategy is used. For almost all test problems the worst (biggest) ratios are achieved when the *breadth first* selection strategy is used.

The total number of simplices (TNS) and the maximum size of candidate list (MCL) at the search tree for different selection strategies are shown in Table 2.11. For $n = 2, 3$-dimensional test problems (TNS) is largest when the *depth first* selection strategy is used. For higher dimensionality $n \geq 4$ the values of (TNS) are very similar for all selection strategies. But the maximal candidate list (MCL) in the search tree varies significantly depending on selection strategies. The best results (the smallest (MCL)) are achieved when the *depth first* selection strategy is used and it is up to 7,000 times smaller than (MCL) with other selection strategies. The maximal candidate list (MCL) is largest when the *breadth first* selection strategy is used. When the *statistical* selection strategy is used (MCL) is up to ~5 times smaller than when the *best first* strategy is used. This explains why execution time is smaller when *statistical* selection strategy is used. This is because insertion and deletion of candidates to/from heap structure depend on the number of elements in the heap.

Let us continue with investigation of selection strategies in the parallel branch-and-bound algorithm with simplicial partitions and aggregate Lipschitz bound $\varphi^1 \mu_2^{2,\infty}$ (2.39). The average values of speedup $\overline{s_p}$ and efficiency of parallelization $\overline{e_p}$ are shown in Table 2.12. For the test problems of dimensionalities $n = 2$ and $n = 3$ the best average efficiency with various numbers of processes p is achieved when the *breadth first* selection strategy is used. The efficiency of

Table 2.11 The total number of simplices and the maximal size of candidate list

#	Best first TNS	MCS	Depth first TNS	MCS	Breadth first TNS	MCS	Statistical TNS	MCS
1	1412	216	9024	14	1430	179	1412	161
2	366	53	396	12	368	30	366	29
3	9684	1843	10622	12	9686	1147	9684	303
4	14	3	14	3	14	4	14	4
5	74	10	74	5	74	16	76	6
6	2480	382	11840	14	2480	283	2480	204
7	24388	5489	24590	13	24388	3603	24402	580
8	678	105	922	11	680	63	678	33
9	43446	10444	43518	14	43446	8508	43446	636
10	2984	483	3210	13	2994	341	2984	93
11	7192	1371	10388	14	7194	1037	7956	210
12	33872	5698	34000	13	33874	7258	33872	1268
13	37466	8559	37560	15	37466	8192	37530	2638
mean	**12620**	**2666**	**14320**	**12**	**12623**	**2359**	**12685**	**474**
14	4710974	901316	4710974	24	4710974	689262	4710974	123184
15	7460	761	12878	20	8046	554	7588	102
16	6291450	1543482	6291450	24	6291450	1572864	6291450	368469
17	2041520	378083	2118242	24	2042846	320538	2041538	154576
18	23954	4660	30044	23	24602	3033	24314	2653
19	34204	6321	217634	24	34204	5534	34204	4780
20	524610	111467	524674	21	524652	76043	524610	11060
mean	**1947739**	**420870**	**1986557**	**23**	**1948111**	**381118**	**1947811**	**94975**
21	10699540	1016395	10701582	39	10699674	1656157	10699540	358063
22	240264	28168	240264	35	240264	27865	240264	5170
23	884808	147320	887536	37	885914	159881	887152	132836
24	1127646	164100	1127648	37	1127646	162612	1127646	144028
25	882342	162790	891788	37	885484	157913	890594	130912
26	1498916	216403	1501446	38	1499432	171915	1499678	16513
27	666916	100877	666918	37	666948	98238	666916	8669
28	538946	57926	550924	39	547820	48453	538946	9126
mean	**2067422**	**236747**	**2071013**	**37**	**2069148**	**310379**	**2068842**	**100665**
29	3267554	496255	3267818	132	3267554	398547	3267554	107605
30	9180776	998386	9180776	133	9180776	977293	9180776	159233
31	10379652	1049544	10405564	138	10413428	784132	10379652	136655
mean	**7609327**	**848062**	**7618053**	**134**	**7620586**	**719991**	**7609327**	**134498**
32	1049160	126600	1049160	728	1049160	126720	1049160	69684
33	3370866	249612	3370866	729	3370866	360473	3370866	142415
mean	**4009784**	**188106**	**4012693**	**729**	**4013537**	**243597**	**4009784**	**106050**

Table 2.12 Average speedup and efficiency of parallelization

n	2 p.		4 p.		8 p.		16 p.		mean $\overline{s_p}$	mean $\overline{e_p}$
	$\overline{s_p}$	$\overline{e_p}$	$\overline{s_p}$	$\overline{e_p}$	$\overline{s_p}$	$\overline{e_p}$	$\overline{s_p}$	$\overline{e_p}$		
					Best first					
2	1.36	0.68	1.95	0.49	2.79	0.35	4.10	0.26	**2.54**	**0.44**
3	1.86	0.93	2.46	0.61	3.64	0.45	5.13	0.32	**3.27**	**0.58**
4	1.95	0.97	3.65	0.91	6.96	0.87	11.14	0.70	**5.98**	**0.86**
5–6	1.87	0.94	3.64	0.91	6.77	0.85	12.95	0.81	**6.31**	**0.88**
mean	**1.76**	**0.88**	**2.93**	**0.73**	**5.04**	**0.63**	**8.33**	**0.52**		
					Depth first					
2	1.61	0.80	1.38	0.35	1.47	0.18	1.39	0.09	**1.47**	**0.36**
3	1.65	0.83	1.92	0.48	2.87	0.36	2.89	0.18	**2.33**	**0.46**
4	1.91	0.96	3.70	0.92	6.80	0.85	9.81	0.61	**5.55**	**0.84**
5–6	1.79	0.89	3.52	0.88	6.69	0.84	12.69	0.79	**6.17**	**0.85**
mean	**1.74**	**0.87**	**2.63**	**0.66**	**4.46**	**0.56**	**6.70**	**0.41**		
					Breadth first					
2	1.35	0.68	2.03	0.51	3.04	0.38	4.99	0.31	**2.85**	**0.47**
3	1.87	0.93	3.17	0.79	5.46	0.68	8.26	0.52	**4.69**	**0.73**
4	1.93	0.96	3.74	0.94	7.01	0.88	9.80	0.61	**5.62**	**0.85**
5–6	1.80	0.90	3.58	0.89	6.75	0.84	13.24	0.83	**6.34**	**0.87**
mean	**1.74**	**0.87**	**3.13**	**0.78**	**5.56**	**0.70**	**9.07**	**0.57**		
					Statistical					
2	1.30	0.65	1.93	0.48	2.96	0.37	4.14	0.26	**2.58**	**0.44**
3	1.83	0.91	2.50	0.63	3.78	0.47	5.02	0.31	**3.28**	**0.58**
4	1.95	0.98	3.72	0.93	7.24	0.90	10.81	0.68	**5.93**	**0.87**
5–6	1.87	0.94	3.62	0.90	6.85	0.86	13.25	0.83	**6.40**	**0.88**
mean	**1.74**	**0.87**	**2.94**	**0.74**	**5.21**	**0.65**	**8.30**	**0.52**		

parallelization is very similar when the *best first* and *statistical* selection strategies are used. The worst efficiency of parallelization for dimensionalities $n = 2$ and $n = 3$ is experienced when the *depth first* selection strategy is used. For higher dimensionalities $n \geq 4$ the average parallel efficiency is similar for all selection strategies. The efficiency of parallelization decreases less with the same number of processes for difficult (higher-dimensional) test problems compared with simpler test problems.

Chapter 3
Simplicial Lipschitz Optimization Without Lipschitz Constant

Global optimization algorithms discussed in the previous chapter, use the global estimate of the Lipschitz constant L given a priori and do not take into account the local information about the behavior of the objective function over every small subregion of $\mathbb{D}$. It has been demonstrated in [74, 116, 126, 134] that estimation of the local Lipschitz constants during the search allows significant acceleration of the global search. Naturally, balancing between the local and global information must be performed in an appropriate way to increase the speed of optimization and avoid the missing of the global solution.

An interesting approach unifying usage of the local and global information during the global search has been proposed in [64]. The algorithm, called DIRECT, was created in the spirit of the Lipschitz optimization, and designed to overcome some of the shortcomings of traditional Lipschitz optimization algorithms. In each iteration of DIRECT algorithm several search points are computed using all possible weights on local versus global search: instead of only one estimate of the Lipschitz constant a set of possible values of L is assumed. Such an approach eliminates the need for "tuning parameters" that set the balance between the local and global search, resulting in an algorithm that is robust and easy to use. DIRECT is especially valuable for engineering optimization problems [1,4,12,17,36,51,81,115]. In such problems the objective and constraint functions are often computed using time-consuming computer simulations, so there is a need to be efficient in the use of function evaluations. While many algorithms address problem features individually, DIRECT is one of a few that address them collectively. However, versatility of DIRECT comes at a cost: the algorithm suffers from a curse of dimensionality that limits applicability to mostly low-dimensional problems ($n \leq 20$).

Almost all papers devoted to DIRECT use hyper-rectangular partitions. However, other types of partitions may be better suited to some problems. This book is devoted to simplicial partitions and therefore in this chapter we build a simplicial DIRECT-type algorithm. However, to make the chapter self contained, we start with an introduction to DIRECT and its modifications. DIRECT was named according to

R. Paulavičius and J. Žilinskas, *Simplicial Global Optimization*,
SpringerBriefs in Optimization, DOI 10.1007/978-1-4614-9093-7_3,
© Remigijus Paulavičius, Julius Žilinskas 2014

one of the primary operations in the procedure—DIviding RECTangles. Similarly we call simplicial algorithm DISIMPL—DIviding SIMPLices [101]. Two versions of DISIMPL are presented in this chapter together with experimental investigation.

3.1 DIRECT Algorithm

Originally published DIRECT algorithm [64] solves problems with bound constraints. The bounds on the variables limit the search to an n-dimensional hyper-rectangular feasible region $\mathbb{D} = [\mathbf{l}, \mathbf{u}] = \{\mathbf{x} \in \mathbb{R}^n : l_i \leq x_i \leq u_i, i = 1, \ldots, n\}$. The real-valued objective function f is supposed Lipschitz-continuous over the feasible region.

DIRECT begins its search for the global minimizer $\mathbf{x}^*$ by fitting the feasible region of the problem $\mathbb{D}$ into the n-dimensional unit hyper-cube

$$\overline{\mathbb{D}} = \{\mathbf{x} \in \mathbb{R}^n : 0 \leq x_i \leq 1, i = 1, \ldots, n\} \tag{3.1}$$

and by evaluating the objective function f in the center point of the unit hyper-cube. DIRECT algorithm works in this normalized space $\overline{\mathbb{D}}$, referring to the original space $\mathbb{D}$ only when evaluating the values of the objective function.

The DIRECT algorithm has two main phases. In the first phase it is necessary to decide which hyper-rectangles are potentially optimal and must be divided in the current iteration, and in the second phase it is necessary to decide how to divide these hyper-rectangles. Figure 3.1 shows the first four iterations in the scaled region $\overline{\mathbb{D}}$ for a two-dimensional Branin test problem. The first row represents the state of DIRECT at the first and the second iterations, and the second row accordingly the third and the fourth iterations. The yellow color highlights potentially optimal rectangles which are selected in the first phase of the current iteration and will be divided in the second phase of the current iteration. The function value obtained in the center of the rectangles are also presented.

At the first iteration, there is only one rectangle—the entire feasible region. The selection of the potential rectangle is therefore trivial and this rectangle is trisected along all its longest dimensions as shown. Such a trisection maintains the property that every sampled center point is at the center of a smaller rectangle. This would not be the case if the rectangles were bisected. At the start of later iterations, the selection process is no longer trivial and will be described in detail.

When DIRECT decides to divide a hyper-rectangle and evaluate the objective function inside it, the hyper-rectangle is referred to as *potentially optimal*. DIRECT algorithm considers a subset of potentially optimal hyper-rectangles satisfying the following definition.

Definition 3.1. Let $\mathbb{S}$ be the set of all hyper-rectangles created by DIRECT after k iterations. Let $\mathbf{c}_i$ denote the center point of the ith hyper-rectangle, and let d_i denote the distance from the center point to one of its corners (it does not matter which corner, since all are equidistant from the center). Let $\varepsilon > 0$ be a positive constant

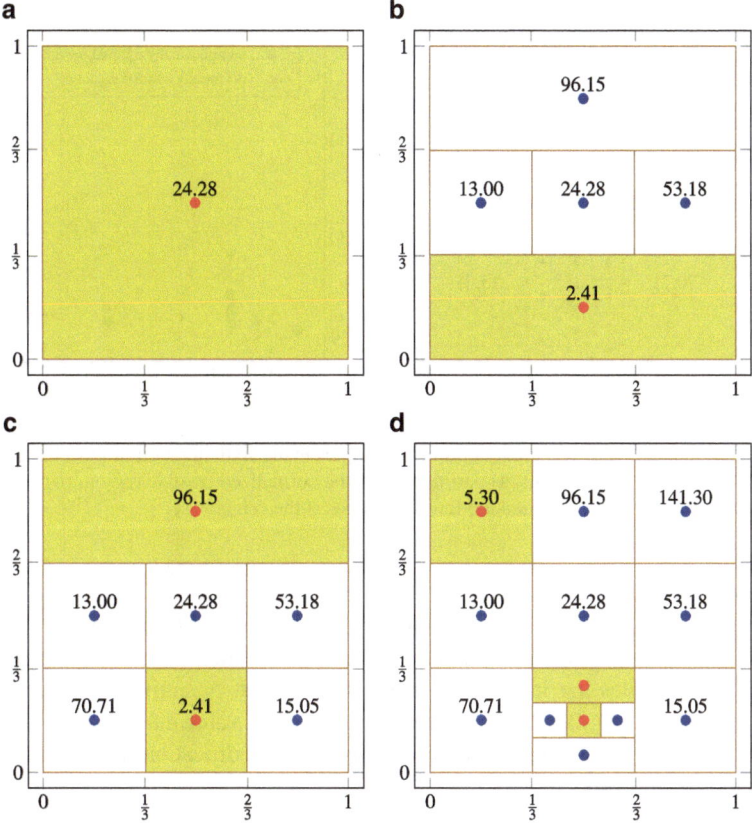

Fig. 3.1 The first iterations of DIRECT algorithm solving the Branin test problem: (**a**) the first iteration, (**b**) the second iteration, (**c**) the third iteration, (**d**) the fourth iteration

and f_{min} be the currently known best function value. A hyper-rectangle $\mathbb{S}_j \in \mathbb{S}$ is said to be potentially optimal if there exists some "rate-of-change" constant $\tilde{L} > 0$ such that

$$f(\mathbf{c}_j) - \tilde{L}d_j \leq f(\mathbf{c}_i) - \tilde{L}d_i, \quad \forall \mathbb{S}_i \in \mathbb{S} \tag{3.2}$$

$$f(\mathbf{c}_j) - \tilde{L}d_j \leq f_{min} - \varepsilon|f_{min}|. \tag{3.3}$$

The parameter ε is used so that $f(\mathbf{c}_j)$ exceeds the current best solution by a nontrivial amount. This condition is needed to prevent the algorithm from becoming too local in its orientation, wasting precious function evaluations in pursuit of extremely small improvements. Experimental investigation [64] has shown that DIRECT is fairly insensitive to the setting of ε, providing good results for values ranging from $10^{-3} \leq \varepsilon \leq 10^{-7}$. A good value for ε is 10^{-4}.

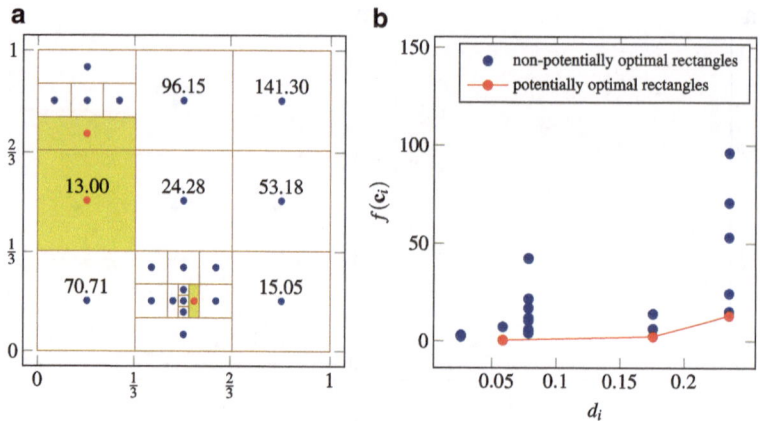

Fig. 3.2 Visualization of potentially optimal rectangles in the sixth iteration of DIRECT algorithm solving the Branin test problem: (**a**) partitioning of the normalized feasible region, (**b**) illustration of hyper-rectangles by the size and the function values at the centers

It was shown in [145] that DIRECT is not strongly homogeneous, i.e., if two objective functions $f(\mathbf{x})$ and $h(\mathbf{x})$ differ only in scales of function values, i.e., $h(\mathbf{x}) = af(\mathbf{x}) + b$ where a and b are constants, then strongly homogeneous algorithm generates the identical sequences of points. A similar observation was shown in [31] that DIRECT with suggested $\varepsilon = 0.0001$ value may have slow asymptotic convergence on poorly scaled problems and a modification that addresses this problem was proposed.

A few observations may be made from definition of potentially optimal rectangles:

- If hyper-rectangle $\mathbb{S}_j$ is potentially optimal, then $f(\mathbf{c}_j) \leq f(\mathbf{c}_i)$ for all hyper-rectangles that are of the same size as $\mathbb{S}_j$ (i.e. $d_j = d_i$).
- If $d_j \geq d_k$ for all k and $f(\mathbf{c}_j) \leq f(\mathbf{c}_i)$ for all hyper-rectangles such that $d_j = d_i$, then hyper-rectangle $\mathbb{S}_j$ is potentially optimal.
- If $d_j \leq d_k$ for all k and $\mathbb{S}_j$ is potentially optimal, then $f(\mathbf{c}_j) = f_{\min}$.

Figure 3.2 illustrates Definition 3.1. Each point in Fig. 3.2b represents a subgroup of hyper-rectangles having equal sizes (horizontal axis) and equal function values at the centers (vertical axis). Equations (3.2)–(3.3) define the set of potentially optimal hyper-rectangles (represented by red points) that correspond to the lower convex hull of the cloud of points (blue).

Potentially optimal hyper-rectangles (the lower convex hull) can be identified in at least two different ways: using modified Graham's scan algorithm [7] or the following lemma [29, 35]:

Lemma 3.1. *Let $\varepsilon > 0$ be a positive constant and let $f_{\min}$ be the current best function value. Let $\mathbb{I}$ be the set of all indices of all hyper-rectangles existing and $j \in \mathbb{I}$ be given. Let*

$$\mathbb{I}_1^j = \{i \in \mathbb{I} : d_i < d_j\}$$
$$\mathbb{I}_2^j = \{i \in \mathbb{I} : d_i > d_j\}$$
$$\mathbb{I}_3^j = \{i \in \mathbb{I} : d_i = d_j\}.$$

Hyper-rectangle $\mathbb{S}_j$ is potentially optimal if

$$f(\mathbf{c}_j) \le f(\mathbf{c}_i), \quad \forall i \in \mathbb{I}_3^j,$$

there exists $\tilde{L} > 0$ such that

$$\max_{i \in \mathbb{I}_1^j} \frac{f(\mathbf{c}_j) - f(\mathbf{c}_i)}{d_j - d_i} \le \tilde{L} \le \min_{i \in \mathbb{I}_2^j} \frac{f(\mathbf{c}_i) - f(\mathbf{c}_j)}{d_i - d_j}$$

and

$$\varepsilon \le \frac{f_{\min} - f(\mathbf{c}_j)}{|f_{\min}|} + \frac{d_j}{|f_{\min}|} \min_{i \in \mathbb{I}_2^j} \frac{f(\mathbf{c}_i) - f(\mathbf{c}_j)}{d_i - d_j}, \quad f_{\min} \ne 0,$$

or

$$f(\mathbf{c}_j) \le d_j \min_{i \in \mathbb{I}_2^j} \frac{f(\mathbf{c}_i) - f(\mathbf{c}_j)}{d_i - d_j}, \quad f_{\min} = 0.$$

Throughout the search, DIRECT will continue to sample in the center of the hyper-cubes and hyper-rectangles. The hyper-rectangles have only one center, regardless of the dimension; therefore DIRECT has a simple and coherent strategy for higher dimensional problems. The division is based on an n-dimensional tri-section. Once a hyper-rectangle has been identified as potentially optimal, DIRECT divides it into smaller hyper-rectangles. The divisions are restricted to only being done along the longest dimension(s) of the hyper-rectangle. This restriction ensures that the rectangles will shrink on every dimension. If the hyper-rectangle is a hyper-cube, then the divisions will be done along all sides.

Let us discuss a general division strategy for hyper-cube and hyper-rectangle in detail. Let $\mathbf{c}$ be the center point of a hyper-cube. The algorithm evaluates the function at the points $\mathbf{c} \pm \delta \mathbf{e}_i$, where δ equals $1/3$ of the side length of the cube and $\mathbf{e}_i$ is the ith Euclidean base vector. DIRECT defines w_i by

$$w_i = \min \{f(\mathbf{c} + \delta \mathbf{e}_i), f(\mathbf{c} - \delta \mathbf{e}_i)\}.$$

The algorithm then divides the hyper-cube in the order given by the w_i, starting with the lowest w_i. DIRECT divides the hyper-cube first perpendicular to the direction with the lowest w_i. Then it divides the remaining volume perpendicular to the direction of the second lowest w_i and so on until the hyper-cube is divided in all directions. This strategy puts $\mathbf{c}$ in the center of a hyper-cube with side length δ, the other $n - 1$ sides will have a length of 3δ. Figure 3.1b illustrates a two-dimensional example: the division of a square for the Branin test problem. There

$$w_1 = \min\{13.00, 53.18\} = 13.00,$$

$$w_2 = \min\{2.41, 96.15\} = 2.41.$$

Therefore we divide first perpendicularly to the x_2-axis, and then in the second step the remaining rectangle is divided perpendicularly to the x_1-axis.

A hyper-rectangle is only divided along its longest sides, which assures the decrease of the maximal side length of the hyper-rectangle. An example can be seen in Fig. 3.1b, c. DIRECT will divide only the highlighted (see Fig. 3.1b) rectangle with $f(\mathbf{c}_i) = 2.41$ (see Fig. 3.1c for the result of subdivision).

DIRECT repeats all main phases of the algorithm until stopping criteria are satisfied. The algorithm is terminated once the percent error (pe) is lower than a given tolerance corresponding to the known global minimum or after the predefined number of iterations. It is also terminated after the maximal allowed number of function evaluations ($M_{max} = 100,000$) have been completed. Let f^* denote the known global minimum and let f_{min} denote the best function value at a current stage of the search, then the percent error is defined by

$$pe = 100\% \times \begin{cases} \frac{f_{min}-f^*}{|f^*|}, & f^* \neq 0, \\ f_{min}, & f^* = 0. \end{cases} \tag{3.4}$$

Note that such a type of stopping criterion is acceptable only when the global minimum f^* or the global minimizer $\mathbf{x}^*$ [123] is known in advance. When such an information is not available, for example in the case of "black-box" optimization, then the most common stopping criterion is the number of executed iterations until a "good" estimate of f^* has been obtained.

3.2 Modifications of DIRECT Algorithm

A slightly modified version of DIRECT was proposed in [63]. The only difference is that in the original DIRECT version a selected hyper-rectangle was trisected not just on a single longest side, but rather on all longest sides. This approach eliminated the need to arbitrarily select a single longest side when there are more than one and, as a result, it added an element of robustness to the algorithm. However, experiments have shown that the robustness benefit is small and that the trisecting on a single longest side accelerates search in higher dimensions.

A locally biased version of DIRECT, named DIRECT-l, was introduced and analyzed in [35, 36]. In the original DIRECT, the size of a hyper-rectangle is measured by the Euclidean distance from its center to a corner. In DIRECT-l, the size of a hyper-rectangle is instead measured by the length of its longest side. Such a measure corresponds to the infinity norm and allows the algorithm to group more hyper-rectangles with the same size. Secondly, in DIRECT-l at most one hyper-rectangle from each group is subdivided, even if there are more than one potentially optimal hyper-rectangle in some groups. This allows reduction of the number of divisions within a group. The results presented in [36] suggest that DIRECT-l should be used for lower dimensional problems, which do not have too many local and global minima. The results also demonstrate the effects of the modifications in DIRECT-l. The convergence occurs in a fewer function evaluations, but can take more iterations. This is due to a fewer hyper-rectangles being chosen in each iteration. However the original DIRECT seems to be the better choice for higher dimensional problems, at least when the termination criterion is based on the difference from the global minimum.

An aggressive version of DIRECT is described in [1]. In this version the authors discard the idea of potentially optimal hyper-rectangles and divide a hyper-rectangle of every size in each iteration. The experiments showed that the aggressive version of DIRECT uses many more function evaluations than the other versions, since the criteria for choosing hyper-rectangles to be divided have been relaxed. Although this approach does not appear to be favorable on the simple test problems, more difficult problems may be more easily solved by this strategy on a large parallel supercomputer [1].

A two-phase approach consisting of explicitly defined global and local phases was proposed in [123]. During the local phase the algorithm tries to explore the subregion around the current best point. The algorithm switches to the global phase when the improvement of at least 1% of the minimal function value is not reached or the hyper-rectangle containing the current best function value becomes the smallest one. Mainly large hyper-rectangles are subdivided in the global phase. It is performed until a function value better than the current best one on at least 1% is obtained. Then the algorithm switches to the local phase during which the obtained new solution is improved locally. The algorithm can switch many times from one phase to another. Thus, the algorithm balances global and local search in a sophisticated way trying to make finding the global minimizers faster. In contrast to all the methods described above that use partitions with evaluation of the objective function once at the central point of each hyper-rectangle, this method works in the framework of the diagonal algorithms evaluating the objective function two times at each hyper rectangle and uses a new efficient partition strategy.

There also exist methods using several Lipschitz constants for the first derivatives together with new efficient one-point-based partitions [72,73].

Finally, availability of modern large scale parallel systems offers potential for solving higher dimensional problems. Unfortunately, the nature of the DIRECT algorithm presents difficulties for an efficient parallel implementation. A few parallel DIRECT implementations are known. First, in [35] a master-slave paradigm was adopted to parallelize only the function evaluations part in the DIRECT algorithm. A fully distributed version of DIRECT is proposed in [140], which was used to solve a 28-dimensional problem on a 256 processor supercomputer. Finally, improved parallel scheme was proposed and experimentally tested on a 2200 processor cluster in [50]. As the authors state, this scheme has been used in the largest application of DIRECT-solving 143-dimensional optimization problems on up to 320 processors in parallel.

3.2.1 Modifications of DIRECT for Problems with Constraints

Many real-world optimization problems contain constraints [12]. Original DIRECT requires bounds on the variables, but does not naturally address other types of linear and nonlinear constraints. In this section, we review some constraint-handling methods from the literature that have been developed for use with DIRECT.

The method proposed by one of the original DIRECT authors [63] involves a nontraditional penalty function and a heuristic for determining the penalty parameters. The key to handling constraints in DIRECT is to work with an auxiliary function that combines information on the objective and constraint functions in a special manner.

A barrier approach assigns a very large function value to infeasible points. Barrier approach is not a good strategy for DIRECT [35] because a hyper-rectangle with a large feasible area, but the infeasible center, will not be explored by DIRECT in a reasonable amount of time.

A traditional exact L1 penalty function approach [32] has been implemented in Matlab software [29]. An L1 penalty approach is a transformation of constrained optimization problem to the form

$$\min \ f(\mathbf{x}) + \sum_{j=1}^{m} r_j g_j(\mathbf{x}) \tag{3.5}$$

$$s.t. \ \mathbf{l} \leq \mathbf{x} \leq \mathbf{u}.$$

where r_j are penalty parameters. The user must provide the values for the penalty parameters in implementation of [29]. Theoretical convergence results for L1 penalty problems are shown in [21, 32], provided a sufficiently large penalty vector, $\mathbf{r}$, is chosen. In some problems, an extremely large penalty parameter is necessary for an algorithm to converge to a feasible point. However, large penalties can make exploring hyper-rectangles near the boundary of the feasible region by DIRECT slow.

A neighborhood assignment strategy (NAS) was proposed in [35]. The strategy assigns values to infeasible points, $\mathbf{x}$, relative to values already found at feasible points in the neighborhood around $\mathbf{x}$. The approach does not need penalty parameters and works with hidden constraints.

A comparison of three different constraint handling approaches (barrier, exact L1 penalty, and NAS) was performed in [30]. The results have shown that DIRECT converges much faster when exact L1 penalty functions are used to handle the constraints. NAS does not use all the available information like constraint violations, thus it is slower. The results of experiments reveal that the barrier functions should not be used with DIRECT. Therefore we use DIRECT with exact L1 penalty functions in our experiments for problems with linear constrains.

3.2.2 SymDIRECT Algorithm

A real function of n variables $f : \mathbb{D} \to \mathbb{R}$, $\mathbb{D} \subseteq \mathbb{R}^n$ is called symmetric if it is unchanged by any permutation of its variables. Global optimization problems with symmetric Lipschitz continuous function occur in various applications [113, 115]. If the symmetric function attains its global minimum in $\mathbf{x}^* = (x_1^*, \ldots, x_n^*) \in \mathbb{D}$ with $x_i^* \neq x_j^*$, then there exist at least $n!$ global minimizers $\tilde{\mathbf{x}}^* = (\tilde{x}_1^*, \ldots, \tilde{x}_n^*) \in \mathbb{D}$, where $(\tilde{x}_1^*, \ldots, \tilde{x}_n^*)$ is any permutation of $(x_1^*, \ldots, x_n^*)$. There are equivalent subregions of the feasible region. To avoid search over equivalent subregions, the constrained global optimization problem

$$f^* = f(\mathbf{x}^*) = \min_{\mathbf{x} \in \Delta} f(\mathbf{x}) \tag{3.6}$$

may be considered, where $\Delta = \{\mathbf{x} \in \mathbb{D} : x_1 \geq \cdots \geq x_n\}$ is the reduced feasible region. Note that the reduced feasible region Δ represents the $n!$th part of the original feasible region $\mathbb{D}$ and in this way the feasible region is significantly reduced.

SymDIRECT [46] is a modification of DIRECT for symmetric Lipschitz continuous functions. SymDIRECT applies DIRECT method to this special situation. A checking is performed for new hyper-rectangles after subdivision. The following situations might occur for a hyper-rectangle $\mathbb{I}$:

1. $\mathbb{I} \subset \Delta$: after subdivision of $\mathbb{I}$ all hyper-rectangles will also be contained in Δ.
2. $\mathbb{I} \cap \Delta = \varnothing$: hyper-rectangle $\mathbb{I}$ cannot contain the solution of reduced (constrained) optimization problem. Hence such hyper-rectangles may be discarded.
3. $\mathbb{I} \cap \Delta \subset \text{bnd } \mathbb{I}$: hyper-rectangle $\mathbb{I}$ touches Δ, but these points will also be on the boundary of other hyper-rectangles. Therefore, we can discard the hyper-rectangle without missing the global minimum.
4. $\mathbb{I} \cap \Delta \neq \mathbb{I}$: some hyper-rectangles resulted during subdivision of $\mathbb{I}$ can be contained in Δ, some may be outside of Δ, and some may be partially in it.

All hyper-rectangles such that $\mathbb{I} \cap \Delta \neq \varnothing$ are candidates for further division and potentially optimal among them should be identified. An efficient method is needed for identification of the hyper-rectangles located at least partially in Δ.

For a two-dimensional symmetric function $f : [0, 1]^2 \to \mathbb{R}$, $\Delta = \{\mathbf{x} = (x_1, x_2) \in [0, 1]^2 : x_1 \geq x_2\}$ all necessary conditions can be noticed easily. Let $\mathbb{I}(\mathbf{c}, (h_1, h_2))$ be a rectangle contained in the unit square $[0, 1]^2$ with the center $\mathbf{c} = (c_1, c_2)$, the half side-lengths h_1, h_2 in the direction of unit vectors $\mathbf{e}_1, \mathbf{e}_2$, and the vertices $\mathbf{v}(v_1, v_2) = (c_1 + \sigma_1 h_1, c_2 + \sigma_2 h_2)$, where $\sigma_1, \sigma_2 \in \{-1, +1\}$.

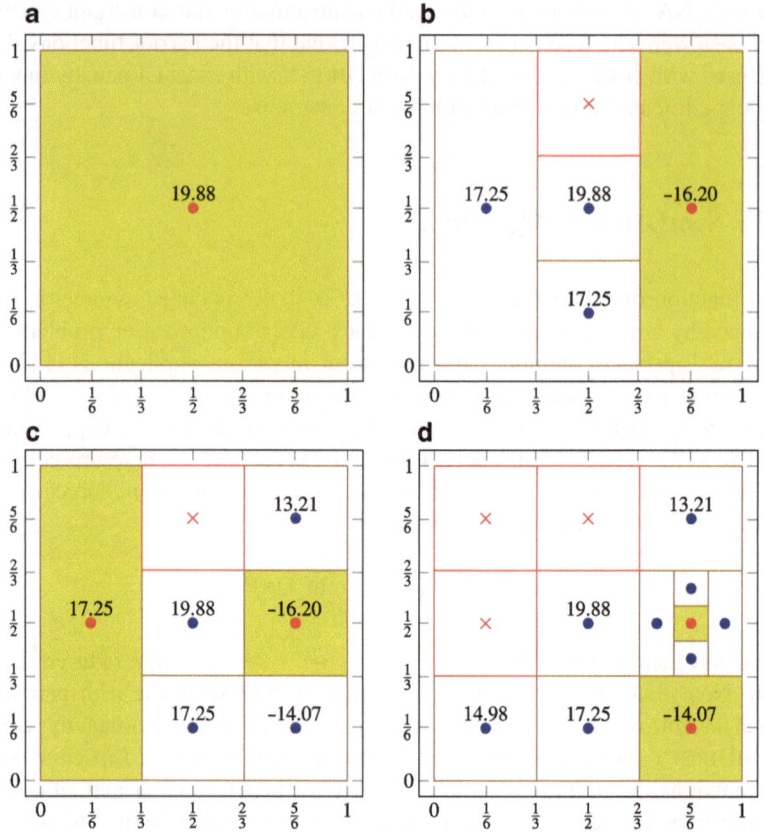

Fig. 3.3 The first iterations of SymDIRECT algorithm solving the symmetric Shubert test problem:
(**a**) the first iteration, (**b**) the second iteration, (**c**) the third iteration, (**d**) the fourth iteration

Note that a point $\mathbf{x} = (x_1, x_2) \in [0, 1]^2$ lies in Δ if and only if $1 \geq x_1 \geq x_2 \geq 0$.
It is easy to see that the whole rectangle $\mathbb{I} \subset \Delta$ if and only if the left upper vertex
$\mathbf{v}(c_1 - h_1, c_2 + h_2) \in \Delta$, and this is fulfilled if and only if there holds

$$c_1 - h_1 \geq c_2 + h_2. \tag{3.7}$$

For example in the third iteration of SymDIRECT (Fig. 3.3c) such rectangles are
with centers $(\frac{1}{2}, \frac{1}{6}), (\frac{5}{6}, \frac{1}{6}), (\frac{5}{6}, \frac{1}{2})$.
A rectangle $\mathbb{I}$ is at least partial in Δ: $\mathbb{I} \cap \Delta \neq \varnothing$ if at least the right lower vertex
$\mathbf{v}(c_1 + h_1, c_2 - h_2) \in \Delta$ if the following condition is fulfilled:

$$c_1 + h_1 \geq c_2 - h_2. \tag{3.8}$$

In the same iteration (see Fig. 3.3c) such rectangles are with centers $(\frac{1}{6}, \frac{1}{2})$, $(\frac{1}{2}, \frac{1}{2})$, $(\frac{5}{6}, \frac{5}{6})$, and $(\frac{1}{2}, \frac{5}{6})$, but the last one only touches Δ and therefore may be discarded.

By checking conditions (3.7) and (3.8) SymDIRECT can accelerate the procedure of dividing rectangles while searching for the global minimum of a symmetric function. Unfortunately such two conditions cannot be generalized for higher dimensional case. If f is a symmetric function, we search only for the global minimizer that lies in the n dimensional simplex $\Delta = \{(x_1, \ldots, x_n) \in [0, 1]^n : x_1 \geq \cdots \geq x_n\}$. A point $\mathbf{x} = (x_1, \ldots, x_n) \in [0, 1]^n$ lies in Δ if and only if $1 \geq x_1 \geq \cdots \geq x_n \geq 0$. The following theorem gives general (any dimensional) conditions by which some hyper-rectangle completely or partially lies in Δ.

Theorem 3.1 ([46]). *Let $\mathbb{I}(\mathbf{c}, h_1, \ldots, h_n)$ be a hyper-rectangle contained in the unit hyper-cube $[0, 1]^n$ with the center $\mathbf{c} = (c_1, \ldots, c_n)$, half side-lengths h_i in the directions of unit vectors $\mathbf{e}_i$, and the vertices $\mathbf{v}(v_1, \ldots, v_n) = (c_1 + \sigma_1 h_1, \ldots, c_n + \sigma_n h_n)$, where $\sigma_1, \ldots, \sigma_n \in \{-1, +1\}$. Then the following holds:*

1. $\mathbb{I} \subset \Delta$ *if and only if the following $(n-1)$ conditions hold:*

$$c_i - h_i \geq c_{i+1} + h_{i+1}, \quad \forall i = 1, \ldots, n-1 \tag{3.9}$$

2. $\mathbb{I} \cap \Delta \neq \emptyset$ *if and only if there exists $\sigma_2, \ldots, \sigma_{n-1} \in \{-1, +1\}$ such that $(2^{n-2}$ possibilities)*

$$c_1 + h_1 \geq c_2 + \sigma_2 h_2 \geq \cdots \geq c_{n-1} + \sigma_{n-1} h_{n-1} \geq c_n - h_n. \tag{3.10}$$

Therefore the global minimum of a symmetric function f will be searched by using SymDIRECT only in $\Delta = \{(x_1, \ldots, x_n) \in [0, 1]^n : x_1 \geq \cdots \geq x_n\}$ dividing a reduced hyper-cube $\overline{\mathbb{D}} = [0, 1]^n$ in the DIRECT algorithm way. Thereby if some hyper-rectangle obtained in the process of dividing a potentially optimal hyper-rectangle falls outside Δ, it can be discarded. If such a hyper-rectangle lies in the region Δ at least partially, it will be divided further. Subsequently, only hyper-rectangles lying at least partially in Δ will be analyzed. This is checked by conditions (3.9)–(3.10).

3.3 DISIMPL Algorithm

In this section we build a simplicial global optimization algorithm [101] based on ideas of DIRECT. By analogy to DIRECT, the new algorithm is called DISIMPL— DIviding SIMPLices. Since a simplex is a polyhedron in n-dimensional space with the minimal number of vertices, it can be advantageous to evaluate the objective function values at its vertices. Therefore we build two versions of the algorithm, one with function values evaluated at the centers and another with function values evaluated at the vertices of the simplices. When the versions must be distinguished, we add a prefix to DISIMPL: -C (centers) and -V (vertices) correspondingly.

Both versions of DISIMPL begin the optimization by fitting the feasible region of the problem $\mathbb{D}$ into the n-dimensional unit hyper-cube

$$\overline{\mathbb{D}} = \{\mathbf{x} \in \mathbb{R}^n : 0 \le x_i \le 1, i = 1, \ldots, n\}$$

similarly as it is done in DIRECT algorithm. DISIMPL continues to work in this normalized space $\overline{\mathbb{D}}$, referring to the original space $\mathbb{D}$ only when evaluating function values.

In order to use simplicial partitions, the next step of DISIMPL algorithm is to cover the normalized feasible region $\overline{\mathbb{D}}$ by simplices. We use combinatorial vertex triangulation (see Sect. 1.3), but other strategies for covering of hyper-cube by simplices may be used as well.

After the initial covering, DISIMPL moves to the next phase of the initial iteration in which it evaluates the objective function values at the centers (DISIMPL-C) or vertices (DISIMPL-V) of the simplices. Therefore, if the initial hyper-cubic feasible region is covered by $n!$ simplices using combinatorial vertex triangulation, 2^n objective function evaluations at the vertices of the hyper-cube are performed using DISIMPL-V or $n!$ evaluations at the centers of the simplices using DISIMPL-C.

Then DISIMPL algorithm begins its main loop of identifying potentially optimal simplices and dividing them appropriately. Figures 3.4–3.7 show partitioned scaled search space $\overline{\mathbb{D}}$ by using DISIMPL-V and DISIMPL-C algorithms on a two-dimensional Branin and a three-dimensional Hartman-3 test problems. The yellow simplices highlight potentially optimal simplices, which are divided in the next phase of the current iteration. DISIMPL-V uses Definition 3.2 and DISIMPL-C uses Definition 3.3 to determine if a simplex is potentially optimal.

Definition 3.2. Let $\mathbb{S}$ be the set of all simplices created by DISIMPL-V after k iterations, $\varepsilon > 0$ be a positive constant, and f_{min} be the currently known best function value. A simplex $\mathbb{S}_j \in \mathbb{S}$ is said to be potentially optimal if there exists some rate-of-change constant $\tilde{L} > 0$ such that

$$\min_{\mathbf{v} \in \mathbb{V}(\mathbb{S}_j)} f(\mathbf{v}) - \tilde{L}\delta_j \le \min_{\mathbf{v} \in \mathbb{V}(\mathbb{S}_i)} f(\mathbf{v}) - \tilde{L}\delta_i, \quad \forall \mathbb{S}_i \in \mathbb{S} \tag{3.11}$$

$$\min_{\mathbf{v} \in \mathbb{V}(\mathbb{S}_j)} f(\mathbf{v}) - \tilde{L}\delta_j \le f_{min} - \varepsilon|f_{min}|. \tag{3.12}$$

Here δ_j denotes a measure for this simplex—the length of its longest edge and $\mathbb{V}(\mathbb{S}_j)$ is the vertex set of the simplex $\mathbb{S}_j$.

Definition 3.3. Let $\mathbb{S}$ be the set of all simplices created by DISIMPL-C after k iterations, $\varepsilon > 0$ be a positive constant, and f_{min} be the currently known best

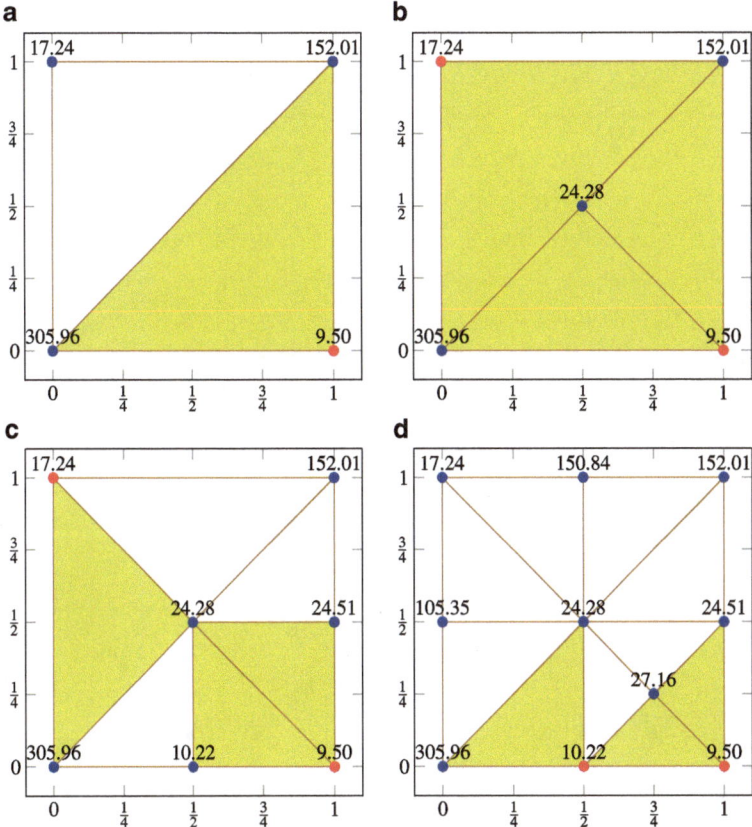

Fig. 3.4 The first iterations of DISIMPL-V algorithm solving Branin test problem: (**a**) the first iteration, (**b**) the second iteration, (**c**) the third iteration, (**d**) the fourth iteration

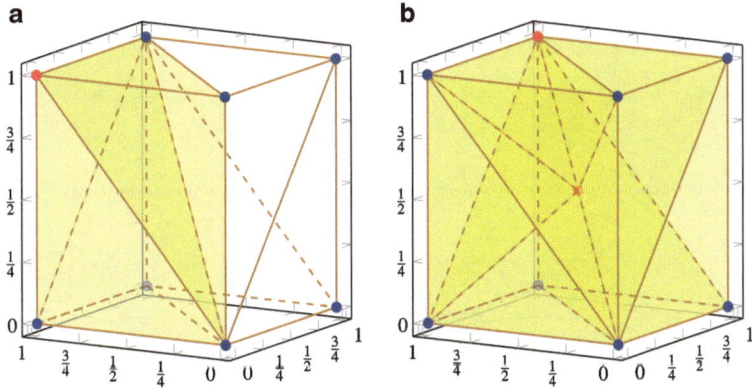

Fig. 3.5 The first iterations of DISIMPL-V algorithm solving three dimensional Hartman-3 test problem: (**a**) the first iteration, (**b**) the second iteration

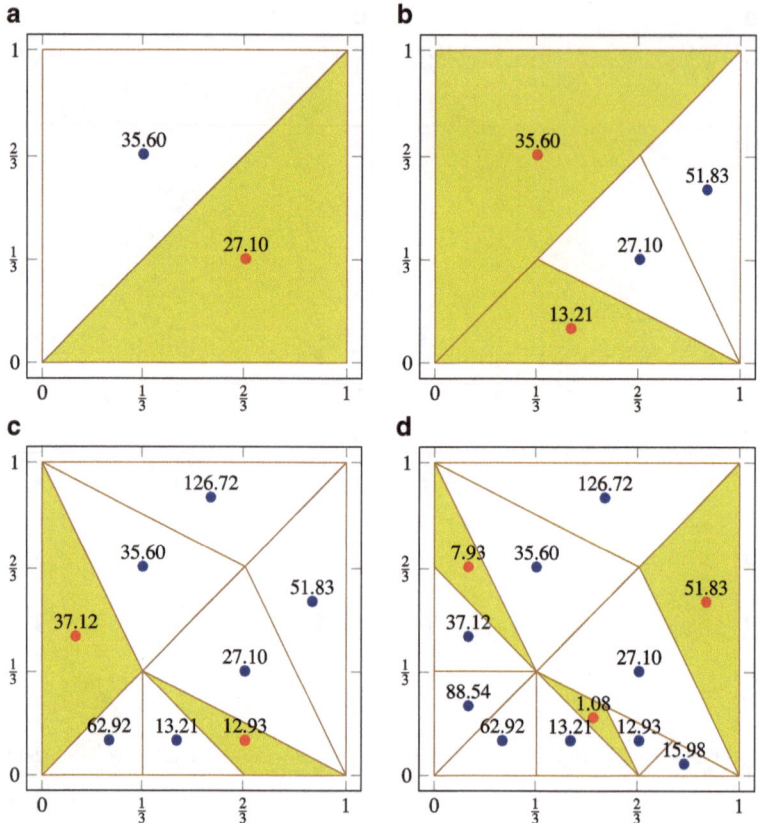

Fig. 3.6 The first iterations of DISIMPL-C algorithm solving Branin test problem: (**a**) the first iteration, (**b**) the second iteration, (**c**) the third iteration, (**d**) the fourth iteration

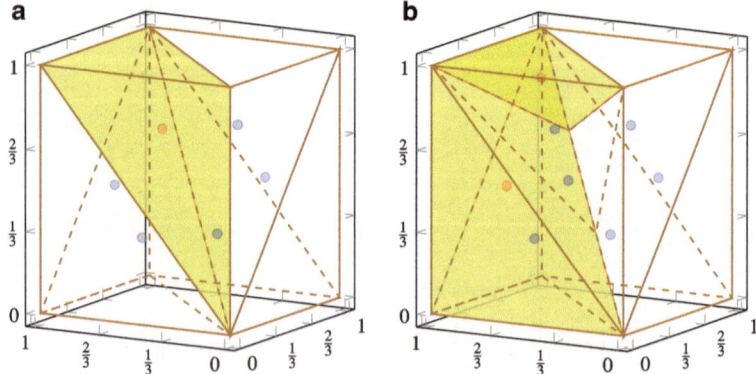

Fig. 3.7 The first two iterations of DISIMPL-C algorithm solving three dimensional Hartman-3 test problem: (**a**) the first iteration, (**b**) the second iteration

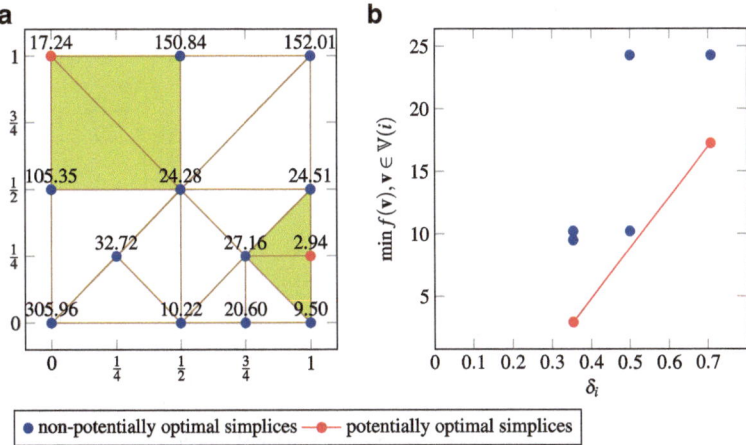

Fig. 3.8 Geometric interpretation of potentially optimal simplices by using DISIMPL-V algorithm on a Branin test problem in fifth iteration: (**a**) partitioning of the normalized feasible region, (**b**) illustration of simplices by the size and the best function value at the vertices

function value. A simplex $\mathbb{S}_j \in \mathbb{S}$ is said to be potentially optimal if there exists some rate-of-change constant $\tilde{L} > 0$ such that

$$f(\mathbf{c}_j) - \tilde{L}\delta_j \leq f(\mathbf{c}_i) - \tilde{L}\delta_i, \quad \forall \mathbb{S}_i \in \mathbb{S} \tag{3.13}$$

$$f(\mathbf{c}_j) - \tilde{L}\delta_j \leq f_{\min} - \varepsilon |f_{\min}|. \tag{3.14}$$

In this definition, $\mathbf{c}_j = \sum\limits_{i=1}^{n+1} \alpha_i v_{j_i}, \alpha_i = 1/(n+1), v_j \in \mathbb{V}(\mathbb{S}_j)$, is the geometric center (centroid) of simplex $\mathbb{S}_j$ and a measure δ_j is the maximal distance from $\mathbf{c}_j$ to the vertices of the simplex.

In both definitions the parameter ε is a "balance parameter" used so that $\min\limits_{v \in \mathbb{V}(\mathbb{S}_j)} f(\mathbf{v})$ or $f(\mathbf{c}_j)$ exceeds the current best function value by a nontrivial amount similarly as in the case of DIRECT.

Figures 3.8b and 3.9b show a geometric interpretation of Definitions 3.2 and 3.3. Each point on the graph represents a simplex in $\mathbb{S}$. Equations (3.11)–(3.12) and (3.13)–(3.14) define the set of potentially optimal simplices that correspond to the lower convex hull of the cloud of points (red points). These simplices are subdivided in the next phase of the current iteration. Once the simplices have been identified as potentially optimal, DISIMPL-V algorithm divides each simplex into two by a hyper-plane passing through the middle point of the longest edge and the vertices which do not belong to the longest edge. DISIMPL-C divides each potentially optimal simplex into three simplices dividing the longest edge in such a way that the center of the divided simplex remains the center of one of the new simplices.

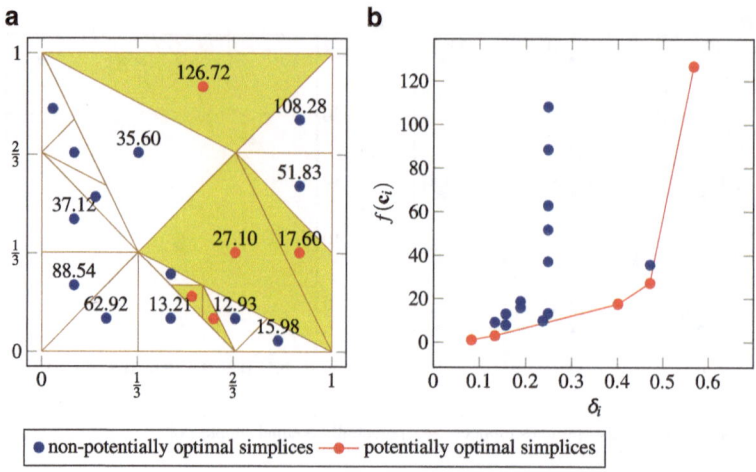

Fig. 3.9 Geometric interpretation of potentially optimal simplices by using DISIMPL-C algorithm on a Branin test problem in fifth iteration: (**a**) partitioning of the normalized feasible region, (**b**) illustration of simplices by the size and the function value at the center

After the subdivision of simplices, the DISIMPL-V algorithm evaluates the objective function $f(\mathbf{x})$ only in new vertices. If the function value at the same point was evaluated in any previous iteration, DISIMPL-V avoids the re-evaluation of the function and reads this value from the previously evaluated function values database. In Fig. 3.8a among 13 points at which the objective function is to be evaluated, $(\frac{1}{2}, \frac{1}{2})$, $(\frac{3}{4}, \frac{1}{4})$ are repeated. Due to the verification, the objective function is evaluated 13 instead of 15 times.

DISIMPL-C algorithm uses trisection strategy and evaluates function values at the geometric centers (centroids) of the new simplices. Since the centroid of one of the new sub-simplices is the same as the centroid of the original simplex, we only need to evaluate the function at two new centroid points. Note that using DISIMPL-C the current number of generated simplices always is equal to the number of function evaluations (trial points).

DISIMPL uses the same termination criteria as DIRECT (see Sect. 3.1). The complete description of DISIMPL algorithm is shown in Algorithm 7. Both versions of DISIMPL have been implemented in C++ and Matlab languages. We use modified Grahams scan algorithm [7] with some tolerance to find potentially optimal simplices. Balanced binary tree is used in DISIMPL-v version to save access time to the vertex data structure.

The convergence properties of DIRECT-type algorithms are widely discussed and investigated [31, 64, 80, 123]. In general, DISIMPL algorithm adopts the ideas of DIRECT using simplicial partitions, therefore convergence properties are very

---

Algorithm 7 DISIMPL

1: Normalize the search space $\mathbb{D}$ to be the unit hyper-cube $\overline{\mathbb{D}}$.
2: Cover $\overline{\mathbb{D}}$ by simplices $\mathbb{S} = \{\mathbb{S}_i | \overline{\mathbb{D}} = \cup \mathbb{S}_i, i = 1, \ldots, n! \}$ using face-to-face vertex triangulation.
3: DISIMPL-V: Evaluate $f(\mathbf{v}_k)$, $k = 1, \ldots, 2^n$. Find f_{min} and pe. Set $m = 2^n$, $t = 0$, $tol = 10^{-2}$.
   DISIMPL-C: Evaluate $f(\mathbf{c}_k)$, $k = 1, \ldots, n!$. Find f_{min} and pe. Set $m = n!$, $t = 0$, $tol = 10^{-2}$.
4: **while** $pe > tol$ and $m < M_{max}$ ($t < T_{max}$) **do**
5:    Identify the set $\mathbb{P} \subset \mathbb{S}$ of potentially optimal simplices.
6:    **while** $\mathbb{P} \neq \emptyset$ **do**
7:       Take $\mathbb{P}_j \in \mathbb{P}$. Set $\mathbb{P} = \mathbb{P} \backslash \{\mathbb{P}_j\}$.
8:       DISIMPL-V: Divide into two new simplices $\mathbb{S}_1$ and $\mathbb{S}_2$. **if** ($\mathbf{v}$ is a new vertex) evaluate $f(\mathbf{v})$. Set $m = m + 1$. Set $\mathbb{S} = \mathbb{S} \cup \{\mathbb{S}_1, \mathbb{S}_2\}$.
         DISIMPL-C: Divide into three new simplices $\mathbb{S}_1$, $\mathbb{S}_2$, and $\mathbb{S}_3$. Evaluate f at new centers. Set $m = m + 2$. Set $\mathbb{S} = \mathbb{S} \cup \{\mathbb{S}_1, \mathbb{S}_2, \mathbb{S}_3\}$.
9:       Update f_{min}.
10:   **end while**
11:   Update pe. Set $t = t + 1$.
12: **end while**

---

similar to that of DIRECT. It follows from (3.11)–(3.12) and (3.13)–(3.14) that the set of potentially optimal simplices always includes at least one simplex from the group of simplices with the biggest diameter δ_{max} (see Figs. 3.8–3.9). If a few of them have the equal measure and achieve the same smallest function value at their vertices (using DISIMPL-v) or at geometric center (using DISIMPL-c), DISIMPL will divide all these simplices. Such simplices will be always potentially optimal with a sufficiently large estimate of the Lipschitz constant $\tilde{L}$. Thus, the simplices with the biggest diameters will be divided through the longest edge in every iteration. Since each group contains only a finite number of simplices, after a sufficiently large number of iterations all simplices of the group with the maximal diameter δ_{max} will be partitioned. Such a procedure will be repeated with a new group of the largest simplices and so on until the largest simplices of the current partition will have the diameter smaller than any $\epsilon > 0$. This shows that for any $\epsilon > 0$ there exists a number of iterations t_ϵ such that $\delta_{max} \leq \epsilon$, if $t \geq t_\epsilon$.

3.3.1 DISIMPL for Lipschitz Optimization Problems with Linear Constraints

In Sect. 3.2.1 we reviewed some constraint-handling methods from the literature that have been developed for using with DIRECT. In the case of DISIMPL a constrained feasible region defined by linear constraints may be covered by simplices. The algorithm remains the same, the only required modification is the initial covering of the feasible region. In Sect. 1.4 we demonstrated that the feasible region defined by linear inequality constraints may be vertex triangulated. To achieve this in DISIMPL

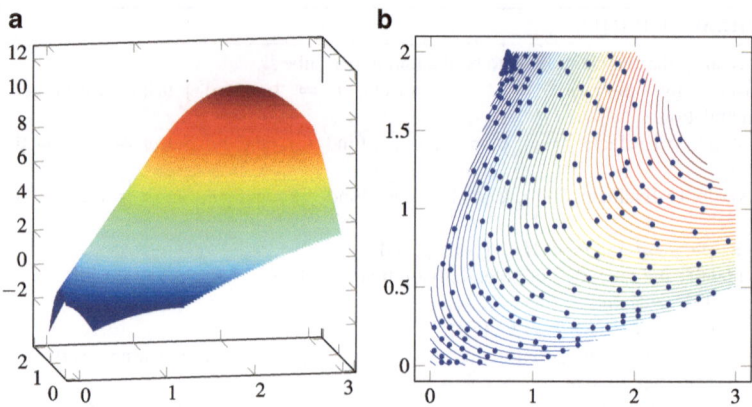

Fig. 3.10 Horst-1 test problem: (**a**) visualization of the objective function, (**b**) scatter plot of sample points of constrained DISIMPL-C

the two step procedure was implemented. First, the vertices of the feasible region are found—they are unique intersection points of all linear constraints. This is done by solving at most $\binom{k}{n}$ n-dimensional linear systems, where k is the number of constraints. Intersection points lying outside the feasible region are rejected. Then we triangulate the feasible region defined by the remaining points (vertices) by using the well-known Delaunay triangulation. After this modified initialization step, DISIMPL algorithm performs the same steps (phases) as described above. This means that no auxiliary computations are required for handling linear constraints except in the initial triangulation of the feasible region.

Let us illustrate the approach by using a two-dimensional numerical test problem Horst-1 from [55], which can be found in Appendix A. The objective function is illustrated in Fig. 3.10a and has several local minima at the vertices of the feasible region. Figure 3.10b presents a scatter plot of DISIMPL-C algorithm solving the Horst-1 test problem. Sampled points clearly cluster around the globally optimal solution $\mathbf{x} = (0.75, 2.0)$. The figure illustrates the situation when DISIMPL-C was terminated with $pe = 0.01$ after 249 function evaluations. Note, that DISIMPL-V terminates with the same percent error after 7 function evaluations, as the optimal solution is obtained on the vertex of the feasible region.

Figure 3.11 shows the first two iterations in scaled constrained region for a Horst-1 test problem. In Fig. 3.12 we also demonstrate the first two iterations of constrained DISIMPL-C on three-dimensional Horst-4 test problem. Even in the first two iterations it can be seen that DISIMPL algorithm samples more around the local and global minimum points. Like for box-constrained problem DISIMPL algorithm performs the same steps (phases) as described above.

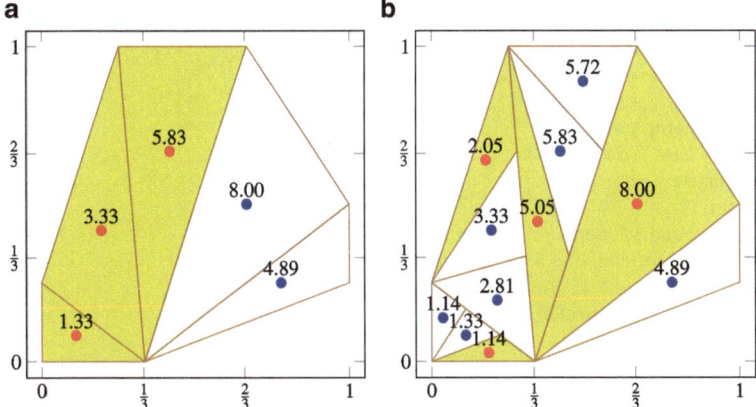

Fig. 3.11 The first two iterations of constrained DISIMPL-C algorithm solving the Horst-1 test problem: (**a**) the first iteration, (**b**) the second iteration

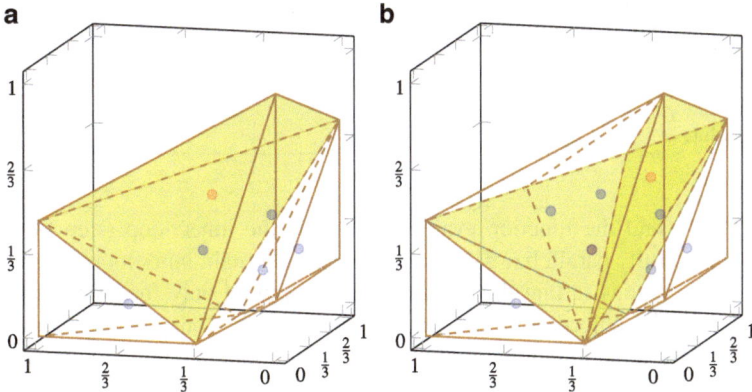

Fig. 3.12 The first two iterations of constrained DISIMPL-C algorithm solving three-dimensional Horst-4 test problem: (**a**) the first iteration, (**b**) the second iteration

3.3.2 Parallel DISIMPL Algorithm

The number of initial simplices grows very fast with the dimension of the problem when combinatorial triangulation is used. Therefore sequential DISIMPL can be effectively used only when the number of variables is small.

In order to solve higher dimension optimization problems parallel DISIMPL version is useful. In this section we describe our implementation for multicore computers with OpenMP. Note that in the sequential DISIMPL algorithm (see Algorithm 7) there are four stages of possible parallelism: covering $\overline{\mathbb{D}}$ by simplices (step 2), function evaluations at covered simplices (step 3), the main inner loop

---

Algorithm 8 Parallel OpenMP version of DISIMPL-V

1: Normalize the search space $\mathbb{D}$ to be the unit hyper-cube $\overline{\mathbb{D}}$.
2: Cover $\overline{\mathbb{D}}$ by simplices $\mathbb{S} = \{\mathbb{S}_i | \overline{\mathbb{D}} = \cup \mathbb{S}_i, \ i = 1, \ldots, n! \}$
3: $\mathbb{V}(\mathbb{S}) = \{\mathbf{v}_j : j = 1, \ldots, 2^n\}$
4: **#pragma omp parallel for**
5: **for** $k = 1$ **to** 2^n **do**
6:      Evaluate $f(\mathbf{v}_k)$
7: **end for**
8: Find f_{min} and pe. Set $m = 2^n, t = 0, tol = 10^{-2}$.
9: **while** $pe > tol$ and $m < M_{\max} \ (t < T_{\max})$ **do**
10:     Identify the set $\mathbb{P} \subset \mathbb{S}$ of potentially optimal simplices.
11:     **#pragma omp parallel for**
12:     **for** $j = 1$ **to** $|\mathbb{P}|$ **do**
13:         Divide $\mathbb{P}_j$ into two new simplices $\mathbb{S}_1$ and $\mathbb{S}_2$.
14:         **if** $\mathbf{v}_j \notin \mathbb{V}(\mathbb{S})$ **then**
15:             Evaluate $f(\mathbf{v}_j)$.
16:             **#pragma omp critical** $(m, \mathbb{V}(\mathbb{S}))$
17:             $\mathbb{V}(\mathbb{S}) = \mathbb{V}(\mathbb{S}) \cup \{\mathbf{v}_j\}$
18:             Set $m = m + 1$.
19:         **end if**
20:         **#pragma omp critical** $(\mathbb{S})$
21:             Set $\mathbb{S} = \mathbb{S} \cup \{\mathbb{S}_1, \mathbb{S}_2\}$.
22:     **end for**
23:     Update f_{min}.
24:     Update pe. Set $t = t + 1$.
25: **end while**

(steps 6–10), and the function evaluations inside the inner loop (step 8). In this approach, we fully parallelize steps 2, 3 and partially main inner loop (steps 6–10). We want to emphasize that this version will be applied only for problems, where evaluation of the objective function at a point is a time-consuming operation (see Sect. 4.3). Objective functions describing real-life applications are often examples of such expensive objective functions.

As the number of simplices for combinatorial triangulation is known in advance, efficient parallel enumeration of all initial simplices may be performed (see Algorithm 3). There is a natural mapping between the integers $0, \ldots, n! - 1$ and permutations of n elements in lexicographical order, when the integers are expressed in factorial form.

Therefore, to perform initial covering of n dimensional hyper-cube by using a given number of threads (NTh), it is sufficient to find correct lexicographical permutation for each of them. After that all processors (threads) can work independently with $n!/NTh$ different permutations.

After parallel combinatorial triangulation, DISIMPL-V performs 2^n and DISIMPL-C performs $n!$ objective function evaluations in parallel. After this part, both versions of DISIMPL begin the main loop of the algorithm. This part in both versions is parallelized partially. The complete parallel DISIMPL algorithms for multi-core computers are presented in Algorithms 8 and 9.

Algorithm 9 Parallel OpenMP version of DISIMPL-C

1: Normalize the search space $\mathbb{D}$ to be the unit hyper-cube $\overline{\mathbb{D}}$.
2: Cover $\overline{\mathbb{D}}$ by simplices $\mathbb{S} = \{\mathbb{S}_i \mid \overline{\mathbb{D}} = \cup \mathbb{S}_i, i = 1, \ldots, n! \}$
3: **#pragma omp parallel for**
4: **for** $k = 1$ **to** $n!$ **do**
5:   Find $\mathbf{c}_k$
6:   Evaluate $f(\mathbf{c}_k)$
7: **end for**
8: Find f_{min} and pe. Set $m = n!, t = 0, tol = 10^{-2}$.
9: **while** $pe > tol$ and $m < M_{\max}$ $(t < T_{\max})$ **do**
10:   Identify the set $\mathbb{P} \subset \mathbb{S}$ of potentially optimal simplices.
11:   **#pragma omp parallel for**
12:   **for** $j = 1$ **to** $|\mathbb{P}|$ **do**
13:     Divide $\mathbb{P}_j$ into three new simplices $\mathbb{S}_1, \mathbb{S}_2$, and $\mathbb{S}_3$.
14:     Find new centers $\{\mathbf{c}_{j1}, \mathbf{c}_{j2}\}$. Evaluate f at new centers.
15:     **#pragma omp critical** $(m, \mathbb{S})$
16:       Set $m = m + 2$. Set $\mathbb{S} = \mathbb{S} \cup \{\mathbb{S}_1, \mathbb{S}_2, \mathbb{S}_3\}$.
17:   **end for**
18:   Update f_{min}.
19:   Update pe. Set $t = t + 1$.
20: **end while**

3.4 Experimental Investigations

In this section we report the results of experimental investigation of DISIMPL. Let us start from numerical results on a class of well-known standard global optimization problems used in [46,64]. In Table 3.1 we summarize the test problems. We give the name of the function, the dimension of the problem, the feasible region over which the problem is defined, the number of global minimizers, and the global minimum (see also in Appendix A).

In experimental investigation we compare the performance of both versions DISIMPL-V and DISIMPL-C with the original DIRECT [64] by using a stand-alone Matlab implementation (*gblSolve.m*) [7] and its modification for symmetric functions SymDIRECT [46] implemented by ourselves. *gblSolve.m* was slightly modified in order to obtain analogous results for test problems reported in [64]. In order to have comparable results $\varepsilon = 0.0001$ was used in all experiments. The efficiency is measured using the number of function evaluations criterion and competition points. For every test problem competition points were calculated as follows: a certain algorithm receives 2 points for the first place (the smallest number of function evaluations), 1 point for the second place, and zero points for the third place. In the case of failure the algorithm gets -1 fine point. Each algorithm is terminated when the percent error (pe) is smaller than 1.0 and 0.01. Failure occurs when the problem takes more than 100,000 function evaluations to satisfy termination criteria.

The experimental results on 12 test problems from [46, 64] are presented in the first part of Table 3.2. The number of function evaluations for the first five test

Table 3.1 Description of test problems used in investigation of DIRECT and its modifications

#	Function name	n	$\mathbb{D}$	No. of global minimizers	Global minimum
1	Branin [23]	2	$[-5, 10] \times [0, 15]$	3	0.398
2	Goldstein-Price [23]	2	$[-2, 2]^2$	1	3.000
3	S-H. Camel B. [143]	2	$[-3, 3] \times [-2, 2]$	2	−1.032
4	Shubert [143]	2	$[-10, 10]^2$	18	−186.831
5	Alolyan [46]	2	$[-1, 1]^2$	1	−1.18519
6	Easom [46]	2	$[-100, 100]^2$	1	3.000
7	Rastrigin [46]	2	$[-5, 6]^2$	2	0
8	Hartman-3 [23]	3	$[0, 1]^3$	1	−3.863
9	Shekel-5 [23]	4	$[0, 10]^4$	1	−10.153
10	Shekel-7 [23]	4	$[0, 10]^4$	1	−10.403
11	Shekel-10 [23]	4	$[0, 10]^4$	1	−10.536
12	Hartman-6 [23]	6	$[0, 1]^6$	1	−3.322
13	Reduced Shekel-5	4	$[0, 6]^4$	1	−10.153
14	Reduced Shekel-7	4	$[0, 6]^4$	1	−10.403
15	Reduced Shekel-10	4	$[0, 6]^4$	1	−10.536

problems are very similar using all three algorithms. However on the Goldstein-Price test problem, for which the global minimizer lies at the boundary of the feasible region, DISIMPL-V with vertex evaluation strategy finishes much faster than DISIMPL-C and DIRECT, which evaluate function values at the central points. For the sixth, unimodal Easom test problem, where the global minimum has a small attraction area relative to the search space, DISIMPL-V also takes significantly less number of function evaluations. For 9–12 test problems DIRECT performs much better comparing with the DISIMPL-V and DISIMPL-C algorithms. However it was shown in [123] that for these functions DIRECT executes a very small number of trials until it generates a point in a neighborhood of the global minimizer. Therefore, such problems are not very suitable for comparison of global optimization methods. The initial feasible region has a significant impact on how quickly an algorithm generates a trial point in the region of attraction of the global minimizer for these problems. Therefore for further investigation, we reduced the feasible region for Shekel problems from $[0, 10]^4$ to $[0, 6]^4$. One would not expect a completely different behavior on such reduced problems. However, in this case (see 13–15 test problems in Table 3.2) we got completely opposite results—the original DIRECT fails for Shekel-5 test problem, and takes more than 30,000 function evaluations for Shekel-7 and Shekel-10 test problems. This happens since DIRECT uses center-point sampling strategy, which has a very slow convergence to locate the global minima on the boundary. Also in this case, we observe that DISIMPL-C finds solution for reduced Shekel-5 test problem after a small number of function evaluations,

Table 3.2 The number of function evaluations (and competition points) on standard test problems

Function name	DISIMPL-V		DISIMPL-C		DIRECT	
	$pe < 1.0$	$pe < 0.01$	$pe < 1.0$	$pe < 0.01$	$pe < 1.0$	$pe < 0.01$
Branin	89 (1)	242 (1)	146 (0)	292 (0)	77 (2)	195 (2)
Goldstein-Price	17 (2)	17 (2)	108 (0)	180 (1)	101 (1)	191 (0)
S.-H. Camel B.	109 (1)	337 (0)	72 (2)	308 (1)	113 (0)	285 (2)
Shubert	4383 (0)	4509 (0)	126 (2)	518 (2)	2883 (1)	2967 (1)
Alolyan	57 (2)	169 (2)	94 (1)	1414 (0)	163 (0)	481 (1)
Easom	1093 (2)	8429 (2)	35578 (0)	40462 (0)	32815 (1)	32845 (1)
Rastrigin	823 (2)	884 (2)	1138 (0)	1306 (0)	843 (1)	969 (1)
Hartman-3	79 (2)	261 (1)	98 (0)	334 (0)	83 (1)	199 (2)
Shekel-5	1999 (1)	2454 (1)	90692 (0)	90948 (0)	103 (2)	155 (2)
Shekel-7	506 (1)	723 (1)	FAIL (−1)	FAIL (−1)	97 (2)	145 (2)
Shekel-10	581 (1)	750 (1)	FAIL (−1)	FAIL (−1)	97 (2)	145 (2)
Hartman-6	454 (1)	6799 (1)	4546 (0)	25334 (0)	213 (2)	571 (2)
R. Shekel-5	883 (1)	1611 (1)	384 (2)	520 (2)	FAIL (−1)	FAIL (−1)
R. Shekel-7	437 (2)	686 (2)	FAIL (−1)	FAIL (−1)	35439 (1)	35529 (1)
R. Shekel-10	386 (2)	675 (2)	FAIL (−1)	FAIL (−1)	34055 (1)	34145 (1)
Total points	(21)	(19)	(3)	(2)	(16)	(19)

however using the original feasible region it failed to find the minimum. However for Shekel-7 and Shekel-10 test problems it also fails to find solution, the same as using the original feasible region.

Let us summarize the results presented in Table 3.2. According to the experimental investigation, the best algorithms for the standard test problems are DISIMPL-V and DIRECT. However DISIMPL-V is the only which successfully solved all test problems. This is mainly influenced by the fact that this algorithm uses more information (the function values at the vertices of the simplices) when selecting potentially optimal simplices as opposite to other two versions, which use less information (the function values at the centers of the subregions) about behavior of the objective function. DISIMPL-C scored less points, but it is clear that failing for Shekel problems had the greatest influence on such bad results. For other test problems this algorithm is also very competitive and this will be obvious from the next results.

We continue by comparing both versions of DISIMPL with modified DIRECT version for symmetric functions (SymDIRECT) [46].

The results of four symmetric functions (4–7 test problems) from Table 3.1 and used in [46] are given in the first part of Table 3.3. We want to note that in these experimental investigations we used our own implementation of SymDIRECT algorithm, as some results presented in [46] are questionable. For example, the global minimum point of Rastrigin test function is at the origin $(0, 0)$ which is the center of the original feasible region $([-5, 5]^2)$. Therefore DIRECT would find the global minimum with one objective function evaluation. However, 379 function evaluations are stated in [46]. To avoid such an accidental success we

Table 3.3 The number of function evaluations (and competition points) on symmetric test problems

Function name	DISIMPL-V		DISIMPL-C		SymDIRECT	
	$pe < 1.0$	$pe < 0.01$	$pe < 1.0$	$pe < 0.01$	$pe < 1.0$	$pe < 0.01$
Shubert	2214 (0)	2275 (0)	63 (2)	259 (2)	1499 (1)	1543 (1)
Alolyan	32 (2)	89 (2)	47 (1)	707 (0)	97 (0)	257 (1)
Easom	564 (2)	4287 (2)	17789 (0)	20231 (0)	16739 (1)	16757 (1)
Rastrigin	442 (2)	474 (2)	569 (0)	653 (0)	481 (1)	559 (1)
$Xcos_{n=2}$	355 (1)	549 (2)	171 (2)	909 (0)	453 (0)	749 (1)
$Xcos_{n=3}$	461 (1)	2113 (1)	337 (2)	1157 (2)	819 (0)	2841 (0)
$Xcos_{n=4}$	144 (2)	144 (2)	411 (1)	959 (1)	2035 (0)	8319 (0)
$Xcos_{n=5}$	1420 (2)	1420 (2)	1599 (1)	2015 (1)	4945 (0)	6629 (0)
$Xcos_{n=6}$	1284 (1)	3832 (1)	409 (2)	837 (2)	10489 (0)	13803 (0)
$Xcos_{n=7}$	2261 (1)	2397 (1)	449 (2)	1213 (2)	11457 (0)	13907 (0)
Total points	(14)	(15)	(13)	(10)	(3)	(5)

use modified feasible region ($[-5, 6]^2$). Improvement of SymDIRECT shown for two-dimensional Alolyan and Easom test problems in [46] is more than 4 and 8 times correspondingly. We do not find any explanation for this improvement, as the search region is reduced less than two times. Our experiments give reasonable results corresponding to reduction of the feasible region by at most $n!$ times.

For most of the test problems the simplicial versions DISIMPL-V and DISIMPL-C give significantly improved results comparing with SymDIRECT. However all these test problems are low dimensional ($n = 2, 3$). Therefore additional investigations were performed solving additional symmetric optimization problem "Xcos" [91] with several variables

$$f(x_1, \ldots, x_n) = \sum_{i=1}^{n} \sum_{j=1}^{n} \left(x_i x_j + 100 \cos x_i \cos x_j \right), \quad x_i \in [-5, 5]^n. \quad (3.15)$$

Testing was conducted for the same cases ($n = 2, \ldots, 7$) as used in [46]. These results are presented in the second part of Table 3.3. Both simplicial versions give significantly better results compared to SymDIRECT algorithm. The difference is especially evident for higher dimensional cases ($n = 5, \ldots, 7$).

The last part of investigations is devoted to problems with linear constraints. We compare DISIMPL versions with a version of DIRECT with exact L1 penalty functions approach, implemented in [29]. The penalty parameters for DIRECT are fixed for each constraint and kept constant during all the iterations. We did some additional testing to see if the numerical results could be improved by perturbing the values of the penalty parameters. We observe completely opposite behavior than described in [139], where no significant differences were observed depending on penalty parameters. In our testing we used three different penalty parameters ($p.p. = 10, p.p. = 10^2, p.p. = 10^3$) for all constraints. The results of seven

Table 3.4 The number of function evaluations (and competition points) solving test problems with linear constrains

Function name	DISIMPL-V		DISIMPL-C		DIRECT-L1	
	$pe < 1.0$	$pe < 0.01$	$pe < 1.0$	$pe < 0.01$	$pe < 1.0$	$pe < 0.01$
$p.p. = 10$					157 (0)	287[a] (0)
Horst-1, $p.p. = 10^2$	7 (2)	7 (2)	123 (1)	249 (1)	2557 (0)	3689 (0)
$p.p. = 10^3$					FAIL (−1)	FAIL (−1)
$p.p. = 10$					101[a] (0)	265[a] (0)
Horst-2, $p.p. = 10^2$	5 (2)	5 (2)	61 (1)	171 (1)	7363 (0)	10829 (0)
$p.p. = 10^3$					25903 (0)	FAIL (−1)
$p.p. = 10$					105 (0)	273 (0)
Horst-3, $p.p. = 10^2$	5 (2)	5 (2)	85 (1)	249 (1)	105 (0)	273 (0)
$p.p. = 10^3$					105 (0)	273 (0)
$p.p. = 10$					2933 (0)	64843 (0)
Horst-4, $p.p. = 10^2$	8 (2)	8 (2)	90 (1)	260 (1)	9553 (0)	FAIL (−1)
$p.p. = 10^3$					10119 (0)	FAIL (−1)
$p.p. = 10$					1411 (0)	3063 (0)
Horst-5, $p.p. = 10^2$	8 (2)	8 (2)	107 (1)	259 (1)	FAIL (−1)	FAIL (−1)
$p.p. = 10^3$					FAIL (−1)	FAIL (−1)
$p.p. = 10$					227[a] (0)	323[a] (0)
Horst-6, $p.p. = 10^2$	10 (2)	10 (2)	102 (1)	284 (1)	1063 (0)	3431 (0)
$p.p. = 10^3$					19873 (0)	FAIL (−1)
$p.p. = 10$					217 (0)	611 (0)
Horst-7, $p.p. = 10^2$	8 (2)	8 (2)	80 (1)	186 (1)	8473 (0)	20819 (0)
$p.p. = 10^3$					FAIL (−1)	FAIL (−1)
$p.p. = 10$					(0)	(0)
Total points, $p.p. = 10^2$	(14)	(14)	(7)	(7)	(−1)	(−2)
$p.p. = 10^3$					(−3)	(−6)

[a] Outside the feasible region

two- and three-dimensional test problems with linear constrains from [55] (see comprehensive description of these problems in Appendix A) are presented in Table 3.4. First, we note big impact of performance by using DIRECT algorithm depending on the penalty parameters. Smaller penalty parameters ($p = 10$) increase the speed of optimization, however in such cases often obtained solution is outside the feasible region. Larger penalty parameters ($p = 10^2$) can assure convergence of the algorithm to a feasible point, however too large penalties ($p = 10^3$) could bias the algorithm away from the hyper-rectangles near the boundary of the feasible region. Of course concepts like "smaller" or "large" penalties are relative and are right only for tested test problems.

However by using DISIMPL we do not have such problems as tuning parameters, because the whole feasible region is triangulated by simplices and further search is performed only in the feasible region. Moreover, we notice that using DISIMPL-C and DISIMPL-V huge improvement is achieved. The number of functions evaluations for all tested problems are better using DISIMPL algorithms. Especially good results are obtained by using DISIMPL-V. This is influenced by the fact that DISIMPL-V evaluates the function values at the vertices of the simplices and therefore for all tested problems the solutions were found after initial triangulation on one of the vertices.

Chapter 4
Applications of Global Optimization Benefiting from Simplicial Partitions

In this chapter we discuss global optimization problems where simplicial partitioning is preferable. Most of the applications discussed here involve global optimization problems with a symmetric objective functions. As it was discussed in Sect. 1.4 the feasible region may be reduced by setting linear constraints in order to avoid equivalent subregions due to the symmetry in the objective function. The resulting constrained feasible region can be covered by simplices and in the case the objective function is invariant to exchange of all variables and the original feasible region is a hyper-cube, the resulting constrained feasible region is a simplex. Therefore such a simplex may be used as a feasible region reducing the hyper-volume by a factor $n!$ times and the numbers of minimizers similarly.

4.1 Global Optimization in Nonlinear Least Squares Regression

One of the problems where simplicial partitioning is useful is global optimization in nonlinear regression [150]. Least squares regression is one of the most frequently used statistical methods. In this section we consider optimization problems occurring in nonlinear least squares regression. Contemporary statistical packages contain algorithms for nonlinear regression including efficient subroutines for local minimization of sums of squared residuals. However, the applied objective functions are often multimodal. Therefore investigation of the applicability of the global optimization algorithms to the problems of nonlinear regression is important.

The optimization problem in nonlinear least squares regression is formulated as

$$\min_{\mathbf{x}\in\mathbb{D}} \sum_{i=1}^{m} (y_i - \varphi(\mathbf{x}, \mathbf{z}_i))^2 = \min_{\mathbf{x}\in\mathbb{D}} f(\mathbf{x}), \qquad (4.1)$$

R. Paulavičius and J. Žilinskas, *Simplicial Global Optimization*,
SpringerBriefs in Optimization, DOI 10.1007/978-1-4614-9093-7_4,
© Remigijus Paulavičius, Julius Žilinskas 2014

where the measurements y_i at the points $\mathbf{z}_i = (z_{1i}, z_{2i}, \ldots, z_{pi})$ should be tuned by the nonlinear function $\varphi(\mathbf{x}, \mathbf{z})$. This is done by searching for the optimal values of the parameters $\mathbf{x}$.

The minimization problem (4.1) seems favorable for application of classical nonlinear programming techniques: the number of variables is small and the objective function is smooth. Indeed, many well-developed nonlinear programming techniques can be applied to find a local minimizer of (4.1). A practical problem could be solved using local optimization algorithms available in various packages if a starting point in the region of attraction of the global minimizer would be known. However, such a starting point is not known, and (4.1) should be considered to be a global optimization problem. This argumentation is supported by the experimental results provided in [75] where it has been shown that the standard algorithms from the statistical packages NCSS, SYSTAT, S-PLUS, SPSS failed to find the global minimizers of considered 14 regression functions for the large percentage of random starting points.

There were a few attempts to attack such optimization problems by randomized heuristics [24, 42, 75, 157], but the considered methods suffer from known general disadvantages of randomized heuristics. The strongest competitors of randomized heuristics for the considered problems seem interval arithmetic based global optimization methods. The objective function (4.1) seems favorable for application of interval optimization methods [48, 92, 105, 153]: the function is defined by a rather simple analytical expression, the number of variables is small, and the formulas of derivatives are available. The extensive testing shows that for such functions interval methods are efficient, and their application in such cases can be recommended [48]. The only difference of the problem (4.1) from the test problems used in the experiments justifying high efficiency of interval methods is the number of summands: the number of summands in (4.1) is typically larger than 10, what can decrease efficiency because of increasing of the dependence. The results of experimental investigation with different nonlinear regression problems have shown that direct application of interval arithmetic based global optimization methods is not promising [148]. Only a few two- and three-dimensional problems have been solved. The success may be improved by rescaling of parameters and analytical reformulation of the problem in the case of presence of linear parameters, what not only makes optimization several times faster, but also enables solution of some problems which are not solved in original formulation during reasonable time. A hybrid global optimization algorithm is more successful [150].

Lipschitz optimization with global constants seems not competitive for such problems since behavior of the objective function is very different in different subsets of the feasible region—the norm of the gradient at the borders of the feasible region may be very huge comparing to other subregions. Adaptive versions of Lipschitz algorithms can be appropriate. In this section we apply Lipschitz optimization without the Lipschitz constants—DIRECT and DISIMPL.

One of the examples of the regression function is

$$\varphi(\mathbf{x}, z) = \exp(x_1 z) + \exp(x_2 z),$$

where there are two nonlinear parameters x_1 and x_2 which are variables in the optimization problem. One can see that this regression function is invariant to exchange of these two parameters. Similarly, the objective function of nonlinear least squares regression is invariant to exchange of the corresponding variables or it can be said that the objective function is symmetric over the line $x_1 = x_2$. Constraints may be set in optimization problem to avoid equivalent subregions: $x_1 \geq x_2$. The resulting search space is a simplex (triangle). In this way the search space is reduced twice. If there is no global minimizer on the line $x_1 = x_2$, the number of global minimizers to find is also reduced twice.

Let us consider another example of the regression function:

$$\varphi(\mathbf{x}, z) = x_1 \exp(x_3 z) + x_2 \exp(x_4 z).$$

The function is invariant to simultaneous exchange of x_1 with x_2 and x_3 with x_4. Therefore, constraints may be set to avoid equivalent subregions of the original feasible region: either $x_1 \geq x_2$ or $x_3 \geq x_4$. The resulting feasible region may be covered by simplices. It can also be seen that this function involves two linear (x_1, x_2) and two nonlinear parameters (x_3, x_4). When the regression function contains linear and nonlinear parameters the optimal values/expressions of the former can be found (similarly to optimal parameters in linear regression) reducing the dimensionality of the problem.

Suppose a general regression function

$$\varphi(\mathbf{x}, \mathbf{u}, \mathbf{z}) = x_0 + \sum_{j=1}^{n} x_j f_j(\mathbf{u}, \mathbf{z}),$$

where $\mathbf{x}$ are linear parameters, $\mathbf{u}$ are nonlinear parameters, and $f_j(\mathbf{u}, \mathbf{z})$ are functions of $\mathbf{u}$ and $\mathbf{z}$, for example $f_j(z) = z$, $f_j(z) = z^2$, $f_j(z) = \exp(z)$, $f_j(u, z) = \exp(uz)$. In such a case the least squares function to minimize is

$$f(\mathbf{x}, \mathbf{u}) = \sum_{i=1}^{N} \left(y_i - x_0 - \sum_{j=1}^{n} x_j f_j(\mathbf{u}, \mathbf{z}_i) \right)^2.$$

Partial derivatives of $f(\mathbf{x}, \mathbf{u})$ with respect to linear parameters $\mathbf{x}$ are

$$\frac{\partial f}{\partial x_0} = \sum_{i=1}^{N} 2 \left(-y_i + x_0 + \sum_{j=1}^{n} x_j f_j(\mathbf{u}, \mathbf{z}_i) \right)$$

$$= 2 \left(-\sum_{i=1}^{N} y_i + x_0 N + \sum_{j=1}^{n} x_j \sum_{i=1}^{N} f_j(\mathbf{u}, \mathbf{z}_i) \right),$$

$$\frac{\partial f}{\partial x_k} = \sum_{i=1}^{N} 2 \left(-y_i f_k(\mathbf{u}, \mathbf{z}_i) + x_0 f_k(\mathbf{u}, \mathbf{z}_i) + \sum_{j=1}^{n} x_j f_j(\mathbf{u}, \mathbf{z}_i) f_k(\mathbf{u}, \mathbf{z}_i) \right)$$

$$= 2 \left(-\sum_{i=1}^{N} y_i f_k(\mathbf{u}, \mathbf{z}_i) + x_0 \sum_{i=1}^{N} f_k(\mathbf{u}, \mathbf{z}_i) + \sum_{j=1}^{n} x_j \sum_{i=1}^{N} f_j(\mathbf{u}, \mathbf{z}_i) f_k(\mathbf{u}, \mathbf{z}_i) \right).$$

The problem to find the optimal values of the linear parameters $\mathbf{x}^*$ such that $\nabla f(\mathbf{x}^*) = \mathbf{0}$ becomes

$$\mathbf{A}\mathbf{x}^* = \mathbf{b},$$

where $\mathbf{A}$ is

$$\mathbf{A} = \begin{pmatrix} N & \sum f_1(\mathbf{u}, \mathbf{z}_i) & \sum f_2(\mathbf{u}, \mathbf{z}_i) & \cdots & \sum f_n(\mathbf{u}, \mathbf{z}_i) \\ \sum f_1(\mathbf{u}, \mathbf{z}_i) & \sum f_1^2(\mathbf{u}, \mathbf{z}_i) & \sum f_1(\mathbf{u}, \mathbf{z}_i) f_2(\mathbf{u}, \mathbf{z}_i) & \cdots & \sum f_1(\mathbf{u}, \mathbf{z}_i) f_n(\mathbf{u}, \mathbf{z}_i) \\ \sum f_2(\mathbf{u}, \mathbf{z}_i) & \sum f_2(\mathbf{u}, \mathbf{z}_i) f_1(\mathbf{u}, \mathbf{z}_i) & \sum f_2^2(\mathbf{u}, \mathbf{z}_i) & \cdots & \sum f_2(\mathbf{u}, \mathbf{z}_i) f_n(\mathbf{u}, \mathbf{z}_i) \\ \vdots & \vdots & \vdots & \ddots & \vdots \\ \sum f_n(\mathbf{u}, \mathbf{z}_i) & \sum f_n(\mathbf{u}, \mathbf{z}_i) f_1(\mathbf{u}, \mathbf{z}_i) & \sum f_n(\mathbf{u}, \mathbf{z}_i) f_2(\mathbf{u}, \mathbf{z}_i) & \cdots & \sum f_n^2(\mathbf{u}, \mathbf{z}_i) \end{pmatrix},$$

and $\mathbf{b}$ is

$$\mathbf{b} = \left(\sum y_i, \ \sum y_i f_1(\mathbf{u}, \mathbf{z}_i), \ \sum y_i f_2(\mathbf{u}, \mathbf{z}_i), \ \ldots, \ \sum y_i f_n(\mathbf{u}, \mathbf{z}_i) \right)^T.$$

Let us remind the example regression function $\varphi(\mathbf{x}, z) = x_1 \exp(x_3 z) + x_2 \exp(x_4 z)$ which contains the linear parameters x_1 and x_2 whose optimal values can be found by solving a system of linear equations

$$\begin{pmatrix} \sum f_1^2(\mathbf{u}, z_i) & \sum f_1(\mathbf{u}, z_i) f_2(\mathbf{u}, z_i) \\ \sum f_2(\mathbf{u}, z_i) f_1(\mathbf{u}, z_i) & \sum f_2^2(\mathbf{u}, z_i) \end{pmatrix} \begin{pmatrix} x_1^* \\ x_2^* \end{pmatrix} = \begin{pmatrix} \sum y_i f_1(\mathbf{u}, z_i) \\ \sum y_i f_2(\mathbf{u}, z_i) \end{pmatrix}.$$

where $f_1(\mathbf{u}, z) = \exp(x_3 z)$, $f_2(\mathbf{u}, z) = \exp(x_4 z)$.

Let us solve such a problem analytically:

$$\begin{pmatrix} \sum f_1^2(\mathbf{u}, z_i) & \sum f_1(\mathbf{u}, z_i) f_2(\mathbf{u}, z_i) \\ \sum f_2(\mathbf{u}, z_i) f_1(\mathbf{u}, z_i) & \sum f_2^2(\mathbf{u}, z_i) \end{pmatrix} \begin{pmatrix} x_1^* \\ x_2^* \end{pmatrix} = \begin{pmatrix} \sum y_i f_1(\mathbf{u}, z_i) \\ \sum y_i f_2(\mathbf{u}, z_i) \end{pmatrix},$$

$$\begin{cases} x_1^* = \frac{\sum y_i f_1(u,z_i) - x_2^* \sum f_1(u,z_i) f_2(u,z_i)}{\sum f_1^2(u,z_i)}, \\ \sum f_2(\mathbf{u}, z_i) f_1(\mathbf{u}, z_i) \frac{\sum y_i f_1(u,z_i) - x_2^* \sum f_1(u,z_i) f_2(u,z_i)}{\sum f_1^2(u,z_i)} + x_2^* \sum f_2^2(\mathbf{u}, z_i) = \sum y_i f_2(\mathbf{u}, z_i), \end{cases}$$

$$\begin{cases} x_2^* = \frac{(\sum y_i f_2(u,z_i))(\sum f_1^2(u,z_i)) - (\sum y_i f_1(u,z_i))(\sum f_1(u,z_i) f_2(u,z_i))}{(\sum f_1^2(u,z_i))(\sum f_2^2(u,z_i)) - (\sum f_1(u,z_i) f_2(u,z_i))^2}, \\ x_1^* = \frac{(\sum y_i f_1(u,z_i))(\sum f_2^2(u,z_i)) - (\sum y_i f_2(u,z_i))(\sum f_1(u,z_i) f_2(u,z_i))}{(\sum f_1^2(u,z_i))(\sum f_2^2(u,z_i)) - (\sum f_1(u,z_i) f_2(u,z_i))^2}. \end{cases}$$

The optimal values of the linear parameters can be expressed substituting $f_1(\mathbf{u}, z) = \exp(x_3 z)$, $f_2(\mathbf{u}, z) = \exp(x_4 z)$:

$$x_1^*(x_3, x_4)$$
$$= \frac{(\sum y_i \exp(x_3 z_i)) \times (\sum \exp(2x_4 z_i)) - (\sum y_i \exp(x_4 z_i)) \times (\sum \exp((x_3+x_4)z_i))}{(\sum \exp(2x_3 z_i)) \times (\sum \exp(2x_4 z_i)) - (\sum \exp((x_3+x_4)z_i))^2},$$

$$x_2^*(x_3, x_4)$$
$$= \frac{(\sum y_i \exp(x_4 z_i)) \times (\sum \exp(2x_3 z_i)) - (\sum y_i \exp(x_3 z_i)) \times (\sum \exp((x_3+x_4)z_i))}{(\sum \exp(2x_3 z_i)) \times (\sum \exp(2x_4 z_i)) - (\sum \exp((x_3+x_4)z_i))^2}$$

with the subsequent reduction of the original regression function with four variables into that with two variables:

$$\varphi(\mathbf{x}, z) = x_1^*(x_3, x_4) \exp(x_3 z) + x_2^*(x_3, x_4) \exp(x_4 z).$$

It can be noted that in this case

$$x_2^*(x_3, x_4) = x_1^*(x_4, x_3),$$

therefore,

$$\varphi(\mathbf{x}, z) = x_1^*(x_3, x_4) \exp(x_3 z) + x_1^*(x_4, x_3) \exp(x_4 z).$$

This function is symmetric: exchange x_3 and x_4 in

$$\varphi_S(\mathbf{x}, z) = x_1^*(x_4, x_3) \exp(x_4 z) + x_1^*(x_3, x_4) \exp(x_3 z)$$

and summands then:

$$\varphi_S(\mathbf{x}, z) = x_1^*(x_3, x_4) \exp(x_3 z) + x_1^*(x_4, x_3) \exp(x_4 z)$$

and we get the same function. Therefore this reduced problem of nonlinear least squares regression is invariant to exchange of the variables corresponding to nonlinear parameters. So symmetries exist in the minimization problem

$$\min_{x_3, x_4} \sum_{i=1}^{N} \left(y_i - x_1^*(x_3, x_4) \exp(x_3 z) - x_1^*(x_4, x_3) \exp(x_4 z) \right)^2.$$

Constraints may be set to avoid equivalent subregions: $x_3 \geq x_4$. The resulting search space is a simplex (triangle).

Some care should be taken if $f_j(\mathbf{u}, \mathbf{z}) = f_k(\mathbf{u}, \mathbf{z})$. In the case $x_3 = x_4$ in our example,

$$f_1(z) = \exp(x_3 z) = \exp(x_4 z) = f_2(z).$$

The optimal value of the linear parameter is undefined:

$$x_1^* = \frac{(\sum y_i \exp(x_3 z_i)) \times (\sum \exp(2x_3 z_i)) - (\sum y_i \exp(x_3 z_i)) \times (\sum \exp(2x_3 z_i))}{(\sum \exp(2x_3 z_i))^2 - (\sum \exp(2x_3 z_i))^2}.$$

This is because the rows of matrix $\mathbf{A}$ become linearly dependent:

$$\begin{pmatrix} \sum f_1^2(z_i) & \sum f_1^2(z_i) \\ \sum f_1^2(z_i) & \sum f_1^2(z_i) \end{pmatrix} \begin{pmatrix} x_1^* \\ x_2^* \end{pmatrix} = \begin{pmatrix} \sum y_i f_1(z_i) \\ \sum y_i f_1(z_i) \end{pmatrix}.$$

In such a case the regression function may be changed to

$$\varphi(\mathbf{x}, z) = x_1^* \exp(x_3 z),$$

where

$$x_1^* = \frac{\sum y_i \exp(x_3 z_i)}{\sum \exp(2x_3 z_i)}.$$

Let us investigate how Lipschitz optimization algorithms without the Lipschitz constants (DIRECT and DISIMPL) solve nonlinear least squares regression problems. We report and comment numerical results on seven nonlinear regression problems from the literature [62, 69, 76, 93]. The problems are reduced to lower dimensional ones if linear parameters exist. Two different feasible regions were used for each problem formulating 14 test problems. Two variants were chosen in order to look at how much influence have the feasible region of these problems. We outline the problems in Table 4.1 where the name and reference, number of variables of optimization problem, regression function, the global minimum f^*, the number of global minimizers $N \mathbf{x}^*$, and the feasible regions are shown.

We compare the performance of two versions of simplicial algorithm DISIMPL-V and DISIMPL-C (see Sect. 3.3) with the original DIRECT [64] by using a stand-alone Matlab implementation (gblSolve) [7] and its modification for symmetric functions SymDIRECT [46] implemented by ourselves. In order to have comparable results $\varepsilon = 0.0001$ was used in all experiments. The efficiency is measured using the number of function evaluations criterion and competition points. The scheme for calculation of competition points is the same as in Sect. 3.4: certain algorithm gets 2 points for the first place (the smallest number of function evaluations); 1 point for the second place and zero points for the third place. In the case of failure the algorithm gets -1 fine point. Each algorithm is terminated when the percent error (pe) is smaller than 1.0 and 0.01. For Test Problems $\# = 9$–14 a more accurate (pe) was used, 10^{-5} and 10^{-7} correspondingly. Failure occurs when the problem takes more than 100,000 function evaluations to satisfy the termination criteria.

Table 4.1 Description of nonlinear least squares regression problems

#	Problem name	n	Regression function	f^*	N	$\mathbf{x}^*$ $\mathbb{D}$
1	Jennrich-Sampson [62]	2	$\exp(x_1z) + \exp(x_2z)$	124.362	1	$[0,1]^2$
2						$[-0.5, 0.5]^2$
3	Militký, Model 2 [69]	2	$\exp(x_1z) + \exp(x_2z)$	8.896×10^{-3}	2	$[-50, 50]^2$
4						$[0, 100]^2$
5	Militký, Model 4 [69]	2	$x_1^* \exp(x_3z) + x_2^* \exp(x_4z)$	3.179×10^{-4}	2	$[-1, 1]^2$
6						$[-2, 0]^2$
7	MGH17 [93]	2	$x_1^* + x_2^* \exp(-x_4z)$	5.465×10^{-5}	2	$[0, 0.05]^2$
8				$+x_3^* \exp(-x_5z)$			$[-0.025, 0.025]^2$
9	Lanczos-1 [76]	3	$x_1^* \exp(-x_2z) + x_3^* \exp(-x_4z)$	1.431×10^{-25}	6	$[0, 10]^3$
10			$+x_5^* \exp(-x_6z)$				$[-5, 5]^3$
11	Lanczos-2 [76]	3	$x_1^* \exp(-x_2z) + x_3^* \exp(-x_4z)$	2.230×10^{-11}	6	$[0, 10]^3$
12			$+x_5^* \exp(-x_6z)$				$[-5, 5]^3$
13	Lanczos-3 [76]	3	$x_1^* \exp(-x_2z) + x_3^* \exp(-x_4z)$	1.612×10^{-8}	6	$[0, 10]^3$
14			$+x_5^* \exp(-x_6z)$				$[-5, 5]^3$

Table 4.2 The number of function evaluations (and competitions points) solving nonlinear regression problems

#	DISIMPL-V		DISIMPL-C		DIRECT	
	$pe < 1.0$	$pe < 0.01$	$pe < 1.0$	$pe < 0.01$	$pe < 1.0$	$pe < 0.01$
1	156 (1)	216 (2)	172 (0)	554 (0)	105 (2)	409 (1)
2	141 (2)	192 (2)	252 (0)	430 (0)	199 (1)	289 (1)
3	28514 (1)	31269 (1)	FAIL (−1)	FAIL (−1)	371 (2)	831 (2)
4	525 (2)	3316 (0)	1110 (0)	1722 (2)	803 (1)	2043 (1)
5	935 (1)	1914 (1)	2456 (0)	3534 (0)	705 (2)	955 (2)
6	939 (1)	1894 (1)	686 (2)	1718 (2)	1577 (0)	2763 (0)
7	288 (1)	1171 (0)	202 (2)	1162 (1)	425 (0)	1025 (2)
8	333 (0)	977 (1)	274 (1)	1006 (0)	201 (2)	815 (2)
9	612 (0)	2439 (2)	258 (1)	5034 (0)	157 (2)	2701 (1)
10	224 (2)	3879 (2)	282 (1)	4050 (1)	407 (0)	5927 (0)
11	612 (0)	2439 (2)	258 (1)	4974 (0)	157 (2)	2701 (1)
12	224 (2)	3879 (2)	282 (1)	4050 (1)	407 (0)	5897 (0)
13	612 (0)	2439 (2)	258 (1)	5034 (0)	157 (2)	2839 (1)
14	224 (2)	4041 (1)	282 (1)	4026 (2)	407 (0)	5633 (0)
Total points	(15)	(19)	(10)	(8)	(16)	(14)

For test problems #=9–14 $pe < 10^{-5}$ and $pe < 10^{-7}$ were used

The number of function evaluations are shown in Table 4.2. The results are similar to those previously obtained using standard test problems in Sect. 3.4, i.e., there is no single algorithm best for all test problems and the results depend on the place of the feasible region. For example, for Militký, Model 2 with centered feasible region DISIMPL-C fails to find the global minimum, but the same Militký, Model 2 with moved feasible region and $pe = 0.01$ DISIMPL-C gives the best

Table 4.3 The number of function evaluations (and competitions points) for nonlinear regression problems taking into account symmetries of the objective functions

#	DISIMPL-V		DISIMPL-C		SymDIRECT	
	$pe < 1.0$	$pe < 0.01$	$pe < 1.0$	$pe < 0.01$	$pe < 1.0$	$pe < 0.01$
1	86 (1)	117 (2)	95 (0)	293 (0)	71 (2)	243 (1)
2	80 (2)	104 (2)	127 (0)	215 (0)	121 (1)	177 (1)
3	14354 (1)	15742 (1)	FAIL (−1)	FAIL (−1)	223 (2)	469 (2)
4	282 (2)	1703 (0)	555 (0)	861 (2)	349 (1)	1099 (1)
5	477 (1)	969 (1)	1257 (0)	1769 (0)	373 (2)	501 (2)
6	479 (1)	958 (1)	347 (2)	869 (2)	893 (0)	1463 (0)
7	151 (1)	598 (0)	101 (2)	581 (1)	229 (0)	531 (2)
8	176 (0)	503 (1)	137 (1)	503 (1)	111 (2)	425 (2)
9	132 (0)	456 (2)	43 (2)	839 (0)	69 (1)	721 (1)
10	57 (1)	713 (1)	47 (2)	675 (2)	199 (0)	1465 (0)
11	132 (0)	456 (2)	43 (2)	829 (0)	57 (1)	709 (1)
12	57 (1)	713 (1)	47 (2)	675 (2)	245 (0)	1393 (0)
13	132 (0)	456 (2)	43 (2)	839 (0)	83 (1)	691 (1)
14	57 (1)	742 (1)	47 (2)	671 (2)	195 (0)	1423 (0)
Total points	(13)	(17)	(16)	(11)	(12)	(14)

For test problems #=9–14 $pe < 10^{-5}$ and $pe < 10^{-7}$ were used

results. According to the experimental investigation, the best algorithm for nonlinear regression problems is DISIMPL-V. It scored the most competition points using greater accuracy $pe < 0.01$. This is mainly influenced by the fact that this algorithm uses more information (the function values at the vertices of the simplices) when selecting potentially optimal simplices as opposite to other two versions, which use less information (the function values at the centers of the subregions) about behavior of the objective function.

All these nonlinear regression problems have symmetries and therefore the numbers of local and global minimizers may be reduced by avoiding searching over equivalent subregions. In the case of DISIMPL this can be done by choosing one instead of $n!$ initial simplices. In order to perform fair comparison, we compare with DIRECT version for symmetric functions (SymDIRECT). The results of nonlinear regression problems are given in Table 4.3. For most of such problems at least one of simplicial versions DISIMPL-V and/or DISIMPL-C give better results comparing with SymDIRECT. The difference is especially evident for $n = 3$ test problems and would increase for higher dimensionality test problems.

Visualization of simplices partitioned by using DISIMPL-V and DISIMPL-C algorithms solving reduced nonlinear least squares regression problem Militký Model 4 [69] is shown in Fig. 4.1a, b. It is clearly visible that the figures are symmetric over the diagonal and only one part should be explored to find the global minimizer. The figures present situation after termination of DISIMPL-V and DISIMPL-C algorithms with $pe = 0.01$. During optimization on the whole (reduced)

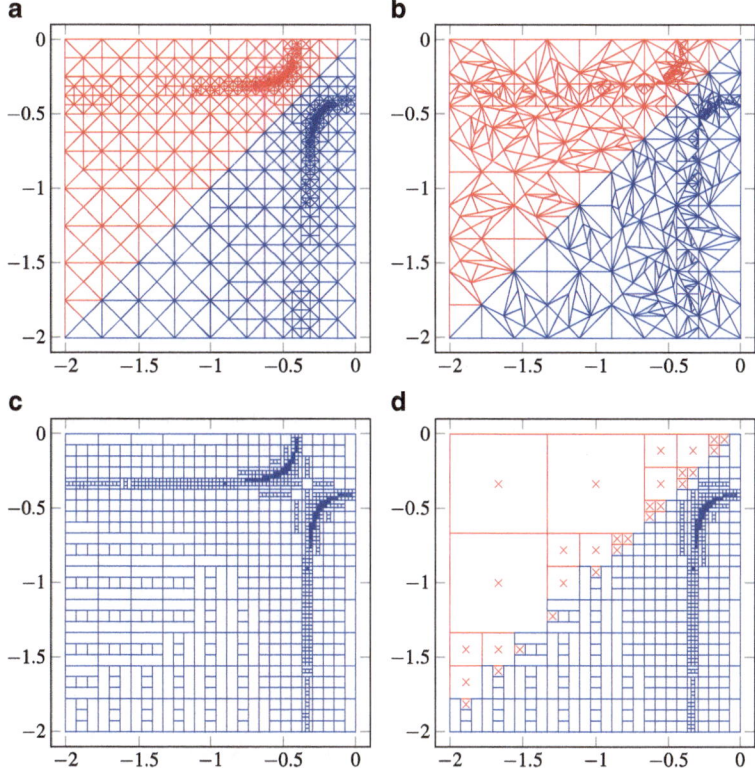

Fig. 4.1 Visualization of simplices and rectangles partitioned solving reduced nonlinear least squares regression problem Militký Model 4: (**a**) DISIMPL-V algorithm, (**b**) DISIMPL-C algorithm, (**c**) DIRECT algorithm, (**d**) SymDIRECT algorithm

search space 1894 (958) function evaluations were performed using DISIMPL-V and 1718 (869) using DISIMPL-C algorithm. As it was noted earlier, the number of simplices using DISIMPL-C is the same as the number of function evaluations. However using DISIMPL-V the number of simplices is larger than the number of function evaluations and the difference becomes larger as the optimization process lasts longer, in this case it is almost two times larger and equal to 3352 (1894).

Visualization of DIRECT and SymDIRECT algorithms solving nonlinear least squares regression problem Militký Model 4 is shown in Fig. 4.1c, d. The presented situation is after DIRECT and SymDIRECT terminate with $pe = 0.01$. DIRECT took 2763 function evaluations optimizing the whole search space and SymDIRECT 1463 on reduced space.

Visualization of simplices partitioned by using DISIMPL-V and DISIMPL-C algorithms solving reduced nonlinear least squares regression problems Jennrich-Sampson and MGH17 are shown in Figs. 4.2 and 4.3. Different shapes of valleys can be seen, but again symmetry is clearly visible.

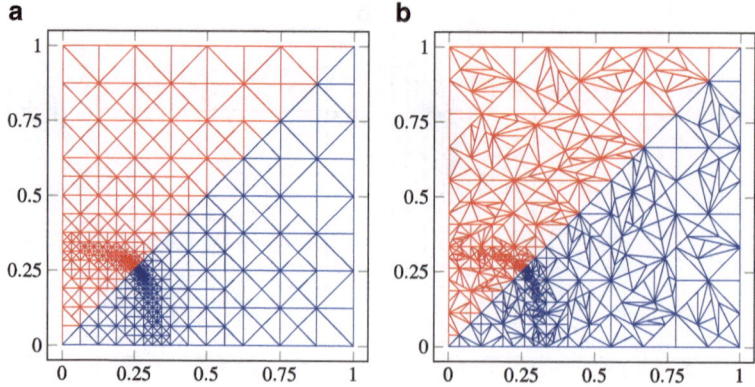

Fig. 4.2 Visualization of simplices partitioned solving reduced nonlinear least squares regression problem Jennrich-Sampson: (**a**) DISIMPL-V algorithm, (**b**) DISIMPL-C algorithm

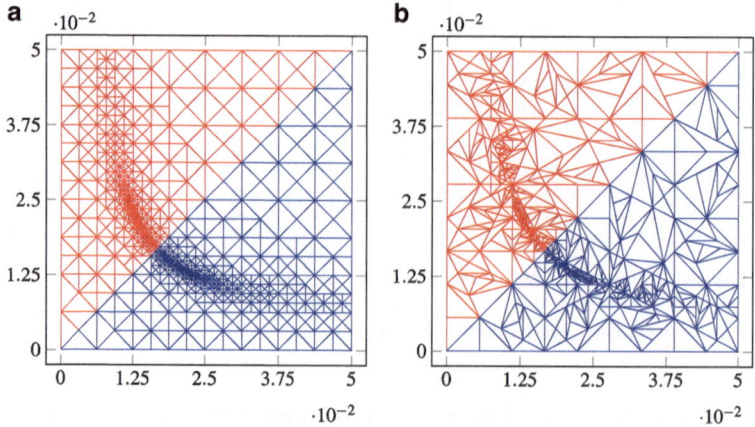

Fig. 4.3 Visualization of simplices partitioned solving reduced nonlinear least squares regression problem MGH17: (**a**) DISIMPL-V algorithm, (**b**) DISIMPL-C algorithm

4.2 Center-Based Clustering Problem for Data Having Only One Feature

Clustering or grouping of data is a well-studied problem in scientific literature. It is important in a wide variety of applications such as biology, information retrieval, text classification, machine learning, business, medicine, psychology, and social sciences [39, 95].

Let $\mathbb{A} = \{a_i \in \mathbb{R} : i = 1, \ldots, m\}$ be a given set of real numbers. The elements of the set $\mathbb{A}$ should be partitioned into n, $1 \leq n \leq m$ nonempty disjoint clusters $\mathbb{A}_1, \ldots, \mathbb{A}_n$, such that

$$\bigcup_{i=1}^{n} \mathbb{A}_i = \mathbb{A}, \quad \mathbb{A}_i \cap \mathbb{A}_j = \emptyset, \quad i \neq j, \quad |\mathbb{A}_j| \geq 1, \quad j = 1, \ldots, n.$$

The partition of the set $\mathbb{A}$ will be denoted by $\Pi(\mathbb{A}) = \{\mathbb{A}_1, \ldots, \mathbb{A}_n\}$.

We can associate the center x_j with each cluster $\mathbb{A}_j$ from the partition $\Pi(\mathbb{A})$, defined by

$$x_j = \arg\min_{x \in \mathbb{I}_j} \sum_{a_i \in \mathbb{A}_j} |x - a_i|,$$

where $\mathbb{I}_j = [\min \mathbb{A}_j, \max \mathbb{A}_j]$. If we define an objective function $f : \mathbb{P}(\mathbb{A}, n) \to [0, +\infty)$ on the set of all partitions $\mathbb{P}(\mathbb{A}, n)$ of set $\mathbb{A}$ containing n clusters by

$$f(\Pi) = \sum_{j=1}^{n} \sum_{a_i \in \mathbb{A}_j} |x_j - a_i|,$$

then we define the optimal partition Π^*, such that

$$f(\Pi^*) = \min_{\Pi \in \mathbb{P}(\mathbb{A}, n)} f(\Pi).$$

Conversely, for a given set of centers $x_1, \ldots, x_n \in \mathbb{R}$ applying the minimal distance condition we can define the partition $\Pi = \{\mathbb{A}_1, \ldots, \mathbb{A}_n\}$ of set $\mathbb{A}$ in the following way:

$$\mathbb{A}_j = \{a \in \mathbb{A} : |x_j - a| \leq |x_i - a|, \quad \forall i = 1, \ldots, n\}, \quad j = 1, \ldots, n,$$

where one has to take care that every element of set $\mathbb{A}$ occurs in one and only one cluster. Therefore the problem of finding the optimal partition of set $\mathbb{A}$ can be reduced to the following optimization problem:

$$\min_{\mathbf{x} \in \mathbb{I}^n} f(\mathbf{x}), \quad f(\mathbf{x}) = \sum_{i=1}^{m} \min_{j=1,\ldots,n} |x_j - a_i|, \tag{4.2}$$

where $\mathbb{I} = [\min \mathbb{A}, \max \mathbb{A}]$. The objective function f is a symmetric Lipschitz continuous function [110, 113], it does not have to be either convex or differentiable and it may have several local and global minima. The number of variables can also be large. Therefore, minimization of this function is a difficult global optimization problem, which is called in the literature as a center-based clustering problem [66, 113].

The most popular algorithm for finding locally optimal partitions is the k-means algorithm [39, 66]. By providing a good initial approximation (see, e.g., [110]), this method can produce acceptable solutions. In case we do not have a good initial approximation, multi-start algorithms with various random initializations are usually recommended [78].

Let us investigate how Lipschitz optimization algorithms without the Lipschitz constants (DIRECT and DISIMPL) solve center-based clustering problems for data having only one feature. The data have been constructed the same way as in [46]. We randomly choose n centers $c_1, \ldots, c_n \in \mathbb{I}_I$, where $\mathbb{I}_I = [0, 100]$. The data set $\mathbb{A}$ containing $m = 100$ real numbers is randomly generated in the following way:

1. Let $i_1, \ldots, i_n$ be randomly generated integers such that $\sum_{s=1}^{n} i_s = m$;
2. In the neighborhood of the center c_s we generate a set $\mathbb{A}_s$, which consists of i_s normally distributed real random numbers from $\mathcal{N}(c_s, 10)$;
3. $\mathbb{A} = \bigcup_{s=1}^{n} \mathbb{A}_s$.

The data set $\mathbb{A} = \{a_1, \ldots, a_m\}$ will be partitioned into $1 \leq n \leq m$ nonempty disjoint clusters $\mathbb{A}_1, \ldots, \mathbb{A}_n$ by solving center-based clustering problem. Let $c_1^*, \ldots, c_n^*$ be the reconstructed centers obtained in the following way:

- Applying SymDIRECT or DISIMPL-V for solving global optimization problem for the constructed data set $\mathbb{A}$ and with some accuracy $\eta > 0$ we obtain $\hat{c}_1, \ldots, \hat{c}_n$;
- Applying the k-means algorithm with the city-block distance for the initial approximation $\hat{c}_1, \ldots, \hat{c}_n$ we obtain reconstructed centers $c_1^*, \ldots, c_n^*$ and the total sum of distances $f(\mathbf{c}^*)$.

The generated data sets $\mathbb{A}$ are shown in Fig. 4.4 where generated random numbers are visualized as blue dots and the obtained centers $c_1^*, \ldots, c_n^*$ as red stars for $n = 2, \ldots, 7$.

Then DISIMPL-C, DISIMPL-V and SymDIRECT algorithms are applied to solve the problem (4.2) with the stopping criterion based on pe (3.4), i.e.

$$\frac{f_{\min} - f(\mathbf{c}^*)}{f(\mathbf{c}^*)} \leq 10^{-2} \, (10^{-4}).$$

Since the objective function is symmetric the search space may be reduced to $x_1 \geq x_2 \geq \cdots \geq x_n$, resulting to a simplex instead of hyper-rectangle $\mathbb{I}^n$. In Fig. 4.5 visualization of simplices partitioned by DISIMPL-V and DISIMPL-C and rectangles partitioned by DIRECT and SymDIRECT solving center-based clustering problem with ($n = 2$) and more precise $pe = 0.001$ is shown. Symmetry over the line $x_1 = x_2$ is clearly visible.

The results of experimental investigation are presented in Table 4.4. For these problems we suppose failure when an algorithm does not finish before 200,000 function evaluations. The results of simplicial DISIMPL-V version are significantly better solving symmetrical clustering problem compared to SymDIRECT algorithm when the dimensionality of the problem is increased ($n \geq 5$). For smaller dimensionality problem ($n \leq 4$) DISIMPL-C is the fastest algorithm. DISIMPL-C

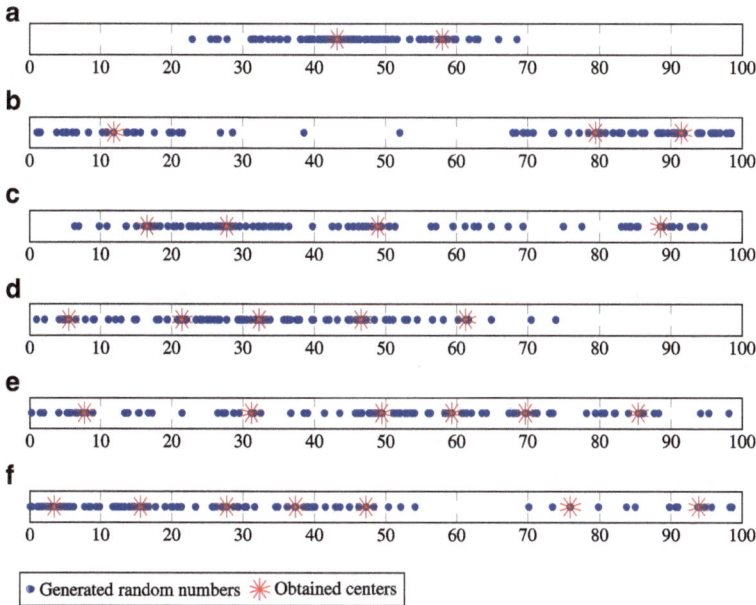

Fig. 4.4 Visualization of generated data set and obtained clusters centers: (**a**) $n = 2$, (**b**) $n = 3$, (**c**) $n = 4$, (**d**) $n = 5$, (**e**) $n = 6$, (**f**) $n = 7$

is also the fastest algorithm for problems with $n = 5$ and $n = 6$ when $pe = 1.0$ was used. However, for a more precise tolerance $pe = 0.01$ it is less efficient than other two compared algorithms.

The distance between adjacent cluster centers cannot be too small. Let us constrain the clustering problem avoiding search over equivalent subregions and distances between two clusters smaller than δ: $x_1 \geq x_2 + \delta, \ldots, x_{n-1} \geq x_n + \delta$. Then the simplicial search space is (see Fig. 4.6)

$$
\mathbb{I} =
\begin{bmatrix}
a + (n-1)\delta & a + (n-2)\delta & \ldots & a + \delta & a \\
b & a + (n-2)\delta & \ldots & a + \delta & a \\
\vdots & & \ddots & & \vdots \\
b & b - \delta & \ldots & b - (n-2)\delta & a \\
b & b - \delta & \ldots & b - (n-2)\delta & b - (n-1)\delta
\end{bmatrix},
$$

where $a = \min \mathbb{A}$ and $b = \max \mathbb{A}$. Such a feasible region can be easily used for simplicial algorithms DISIMPL-V and DISIMPL-C, but SymDIRECT must be modified to account for such constraints. The results of experimental investigation with $\delta = 9$ are shown in Table 4.5. As it can be seen from the results, the reduction of the feasible region usually leads to improved results. However, sometimes it can significantly worsen the results as for DISIMPL-C algorithm solving $Cluster_{n=4}$

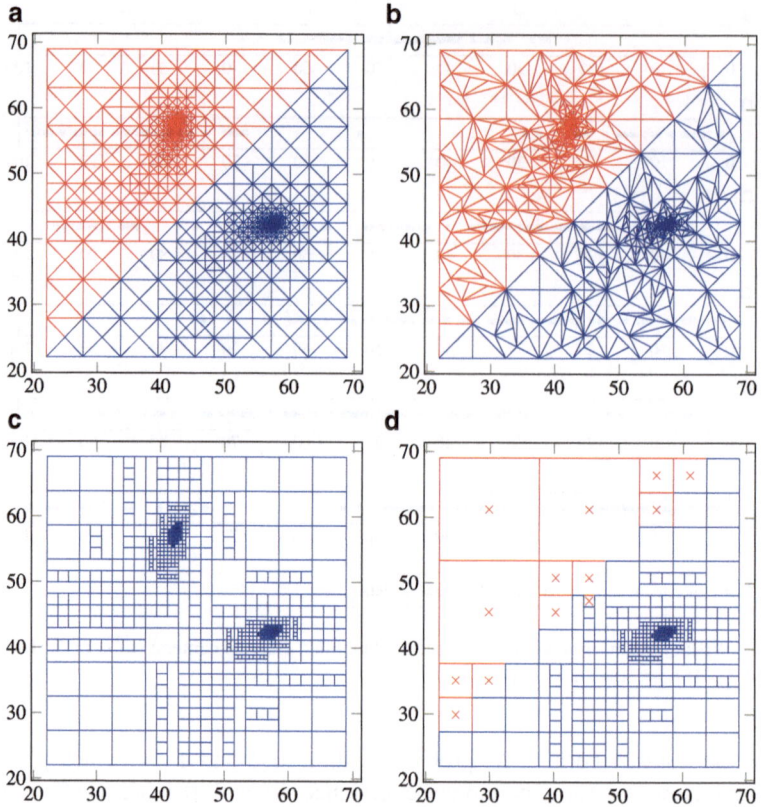

Fig. 4.5 Visualization of simplices and rectangles partitioned solving center-based clustering problem ($n = 2$): (**a**) DISIMPL-V algorithm, (**b**) DISIMPL-C algorithm, (**c**) DIRECT algorithm, (**d**) SymDIRECT algorithm

Table 4.4 The number of function evaluations (and competitions points) solving center-based clustering problems

#	DISIMPL-V		DISIMPL-C		SymDIRECT	
	$pe < 1.0$	$pe < 0.01$	$pe < 1.0$	$pe < 0.01$	$pe < 1.0$	$pe < 0.01$
$Cluster_{n=2}$	37 (0)	165 (0)	27 (2)	95 (2)	29 (1)	121 (1)
$Cluster_{n=3}$	207 (1)	512 (0)	77 (2)	351 (2)	213 (0)	417 (1)
$Cluster_{n=4}$	865 (0)	4358 (0)	659 (1)	887 (2)	605 (2)	1047 (1)
$Cluster_{n=5}$	2558 (0)	4173 (2)	985 (2)	34569 (0)	2262 (1)	5175 (1)
$Cluster_{n=6}$	4748 (1)	9791 (2)	2019 (2)	116305 (0)	8001 (0)	25537 (1)
$Cluster_{n=7}$	10489 (2)	45418 (2)	92901 (0)	FAIL (−1)	30139 (1)	141849 (1)
Total points	(4)	(6)	(9)	(5)	(5)	(6)

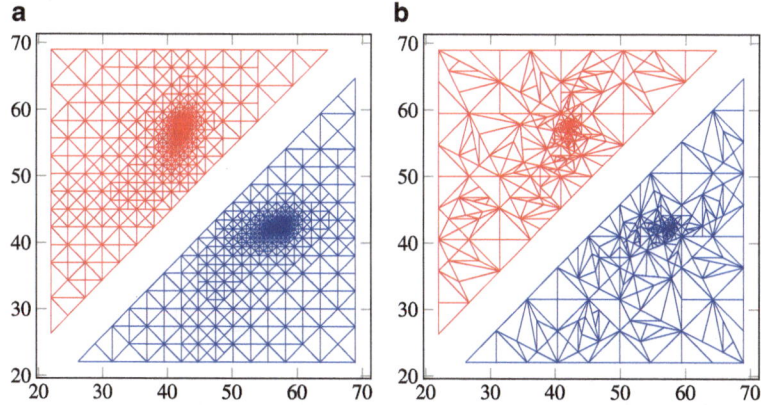

Fig. 4.6 Visualization of simplices partitioned solving center-based clustering problem ($n = 2$) on reduced search space: (**a**) DISIMPL-V algorithm, (**b**) DISIMPL-C algorithm

Table 4.5 The number of function evaluations (and competitions points) solving center-based clustering problems on reduced search space

	DISIMPL-V		DISIMPL-C		SymDIRECT	
#	$pe < 1.0$	$pe < 0.01$	$pe < 1.0$	$pe < 0.01$	$pe < 1.0$	$pe < 0.01$
$Cluster_{n=2}$	11 (2)	115 (2)	43 (0)	127 (0)	29 (1)	121 (1)
$Cluster_{n=3}$	100 (1)	474 (0)	55 (2)	267 (2)	213 (0)	417 (1)
$Cluster_{n=4}$	129 (2)	3127 (0)	283 (1)	2111 (1)	605 (2)	1047 (2)
$Cluster_{n=5}$	438 (1)	12168 (1)	427 (2)	20817 (0)	2262 (1)	5175 (2)
$Cluster_{n=6}$	1464 (2)	4217 (2)	1393 (1)	148934 (1)	8001 (0)	25537 (1)
$Cluster_{n=7}$	607 (2)	26463 (2)	76301 (1)	FAIL (−1)	30139 (1)	141849 (1)
Total points	(10)	(7)	(7)	(3)	(5)	(8)

problem and DISIMPL-V algorithm solving $Cluster_{n=5}$ problem with $pe = 0.01$. And even a small delta parameter reduction from $\delta = 9$ to $\delta = 8$, increasing the feasible region, may lead to much better results: for DISIMPL-V algorithm solving $Cluster_{n=5}$ with the same $pe = 0.01$ and reduction parameter $\delta = 8$ the total number of function evaluation is reduced to 2061. Therefore, the issue of selecting optimal δ parameter is still open and requires further investigation.

4.3   Pile Placement in Grillage-Type Foundations

Grillage-type foundations are the most conventional and effective scheme of foundations, especially in the case of weak grounds. Grillage consists of separate beams, which are supported by piles or reside on other beams. As piles may reach length of 10 m, reduction of the number of piles leads to substantial savings. The

Table 4.6 Characteristics of problems of grillage-type foundations

Problem (#)	n	l	R_{allw}	R_{ideal}
1	25	172.9	325	307.47
2	18	52.9	110	104.12
3	31	84.1	105	101.85
4	31	84.9	105	101.24
5	30	63.9	100	97.51
6	37	80.1	100	97.53
7	23	129.1	300	287.35
8	34	137.9	250	236.28
9	17	97.6	250	244.71
10	55	315.61	350	349.05

optimal scheme of grillage should possess the minimum possible number of piles. Theoretically, reactive forces in all piles should approach the limit magnitudes of reactions for them [6]. This goal can be achieved by choosing appropriate pile positions. Therefore, the piles should be positioned minimizing the largest difference between the reactive forces and the limit magnitudes of reactions.

A designer may arrive at an acceptable pile placement scheme by engineering tests algorithms. However, obtaining the optimal schemes is likely only in the case of simple geometries, simple loadings, and the limited number of design parameters. Practically, this is difficult to achieve for grillages of complex geometries. To be on the safer side, the number of piles in design schemes is usually overestimated.

Optimization problems of pile placement may be approached using global optimization [15]. These optimization problems are "black box" type: the values of the objective function are evaluated by an independent package which models reactive forces in the grillage using finite element method. The number of piles is n, usually $n \geq 10$.

We perform investigation of DISIMPL with the grillage problem instances from [5]. The modeling package and problem instances are available at http://www.gridglobopt.vgtu.lt/sijynai. The characteristics of the problems are given in Table 4.6. R_{allw} denotes the allowable reactive force at the piles. In all problems it is equal for all piles of grillage. The number of piles n is obtained by dividing the sum of loadings by the allowable reactive force of piles—the number of piles cannot be less than this number. R_{ideal} denotes the ideal value. The ideal solution is when reactive forces at all piles are equal. In this case the reactive forces are equal to the sum of loadings divided by the number of piles. In all these problems the proportion between the total loading and the allowable reaction is such that the engineering solution requires achieving almost the ideal solution. This is the main reason why these problems are difficult to solve.

The position of a pile is defined by a real number, which is mapped to a two-dimensional position by the modeling package. Illustration for Problem 2 is given in Fig. 4.7. Beams are shown by line segments of different colors. Piles are illustrated by dots.

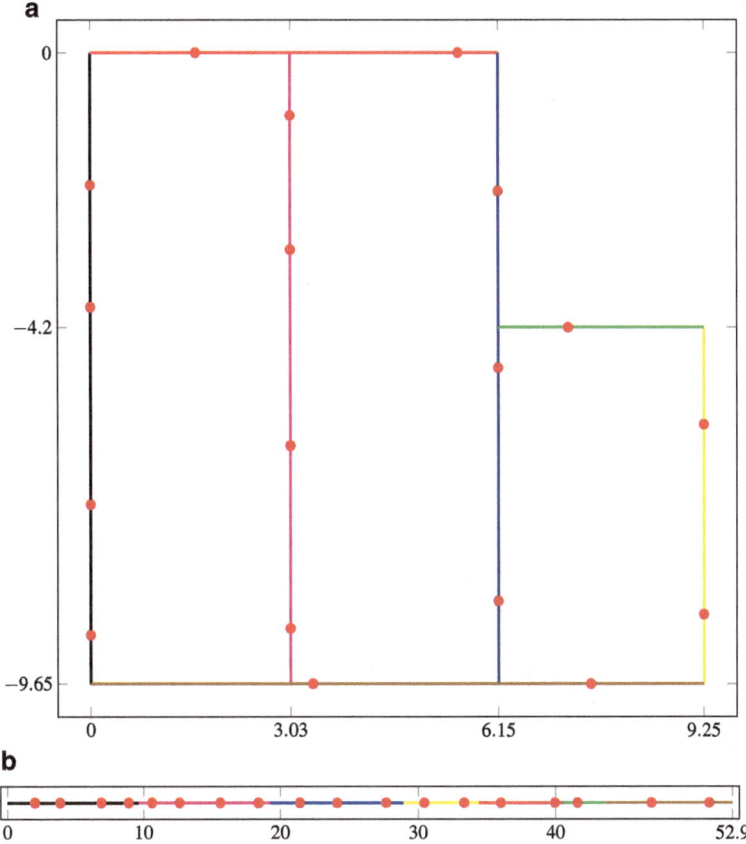

Fig. 4.7 Illustration of Problem 2 with $n = 18$ piles: (**a**) geometry, (**b**) one-dimensional representation

Possible values for variables representing positions of piles are from zero to the sum of length of all beams l. The feasible region of the problems is $[0, l]^n$.

If characteristics of all piles are equal, interchanging them does not change the value of the objective function. The problem may be constrained to avoid equivalent subregions of the feasible region: $x_1 \geq x_2 \geq \cdots \geq x_n$. In this case the search space is a simplex

$$\mathbb{I} = \begin{bmatrix} 0 & 0 & \dots & 0 & 0 \\ l & 0 & \dots & 0 & 0 \\ \vdots & & \ddots & & \vdots \\ l & l & \dots & l & 0 \\ l & l & \dots & l & l \end{bmatrix}.$$

The search space and the numbers of local and global minimizers are reduced by a factor $n!$ times with respect to the original feasible region.

Two piles may not be at the same position. Let us define the minimal distance between two piles by δ. The problem may be constrained to avoid search over equivalent subregions and coincidence of the piles:

$$x_1 \geq x_2 + \delta, \ldots, x_{n-1} \geq x_n + \delta.$$

In this case the search space is the simplex

$$
\mathbb{I} =
\begin{bmatrix}
(n-1)\delta & (n-2)\delta & \ldots & \delta & 0 \\
l & (n-2)\delta & \ldots & \delta & 0 \\
\vdots & & \ddots & & \vdots \\
l & l-\delta & \ldots l-(n-2)\delta & 0 \\
l & l-\delta & \ldots l-(n-2)\delta & l-(n-1)\delta
\end{bmatrix}.
$$

The search space and the numbers of local and global minimizers are reduced approximately $n!$ times with respect to the original feasible region. Moreover in this case there are no unfeasible subregions in the search space where piles coincide.

We compare the results of DISIMPL-C algorithm with the results of several algorithms given in [5]:

- Random search (RS);
- Modified random search (MRS);
- Simulated annealing (SA);
- Genetic algorithm (GA);
- Simplex method (SM);
- Variable metric method (VM);
- NEWUOA algorithm.

Investigated grillage problems are of small-to-medium scale, requiring from 18 to 55 piles. We want to emphasize that according to the authors of DIRECT [63], the versatility of the algorithm limits it to low-dimensional problems ($n \leq 20$). Therefore these problems are definitely challenging to the same class DISIMPL-C algorithm. Also notice that DISIMPL-C is the only deterministic algorithm used to solve these problems. The typical grillage together with obtained pile positions is shown in Fig. 4.7.

Tables 4.7 and 4.8 present the average results when RS (Table 4.7) or MRS (Table 4.8) was used for search and initialization for the comparable algorithms. Bold numbers represent the best results. To make a fair comparison, the total number of objective function evaluations for each algorithm is the same: 5,000. DISIMPL-C is deterministic and therefore it is run only once for each problem. From these tables we can see that the deterministic DISIMPL-C is very competitive, and most often was the best of all the compared algorithm when RS was used for search and initialization.

Table 4.7 Average of the best values found in 28 runs when RS is used for search and initialization

#	R_{ideal}	RS	SA	GA	SM	VM	NEWUOA	DISIMPL-C
1	307.47	593.33	463.08	490.01	633.77	547.63	440.31	**394.70**
2	104.12	153.07	119.74	119.60	172.79	152.38	118.68	**111.55**
3	101.85	258.45	**134.59**	142.24	274.06	230.73	139.88	140.68
4	101.24	265.41	139.63	149.87	264.42	251.90	**136.43**	137.15
5	97.51	318.16	132.52	142.53	310.40	290.67	128.79	**112.04**
6	97.53	460.31	162.09	175.90	460.16	435.57	188.60	**150.50**
7	287.35	472.74	372.63	380.48	507.99	450.81	369.03	**357.20**
8	236.28	695.60	**455.63**	479.87	721.70	655.56	467.22	504.62
9	244.71	402.17	307.55	321.27	427.50	385.72	296.27	**283.13**
10	349.05	1321.48	806.46	890.51	1319.49	1163.32	771.26	**659.55**

Table 4.8 Average of the best values found in 28 runs when MRS is used for search and initialization

#	R_{ideal}	MRS	SA	GA	SM	VM	NEWUOA	DISIMPL-C
1	307.47	470.90	**371.55**	405.36	486.50	454.01	394.92	394.70
2	104.12	125.71	**109.10**	112.49	131.75	127.64	113.35	111.55
3	101.85	144.46	**119.06**	124.48	153.82	143.60	119.80	140.68
4	101.24	141.18	117.10	123.72	147.46	139.82	**116.81**	137.15
5	97.51	126.08	**106.25**	112.23	133.28	128.13	108.68	112.04
6	97.53	160.18	132.22	144.16	172.38	160.85	**132.07**	150.50
7	287.35	379.80	**314.11**	332.11	402.36	382.45	330.16	357.20
8	236.28	494.83	413.57	444.88	520.98	491.25	**385.99**	504.62
9	244.71	343.91	**281.12**	294.62	369.06	334.90	292.28	283.13
10	349.05	702.53	562.79	636.87	759.33	705.57	**559.88**	659.55

As the evaluation of the objective function at a point is a most time-consuming operation solving this real-life application, both parallel DISIMPL OpenMP versions were tested. Experiments were performed using the following hardware and software system:

- Quad-core Intel Core i7 Sandy Bridge with 6 MB on-chip L3 cache (2.4 GHz), 1, 2 or 4 cores are used for the parallel version of the algorithm;
- 8GB RAM;
- OS X 10.8.4;
- Xcode 4.6.3;
- LLVM GCC 4.2 Compiler.

The parallelization was measured using speedup and efficiency criteria. The plots of the results on pile placement problem are given in Fig. 4.8. The results show that efficiency of parallelization is good. It can also be seen that the efficiency of parallelization is better for larger problems. The figures reveal that efficiency of parallelization is very similar in both DISIMPL versions.

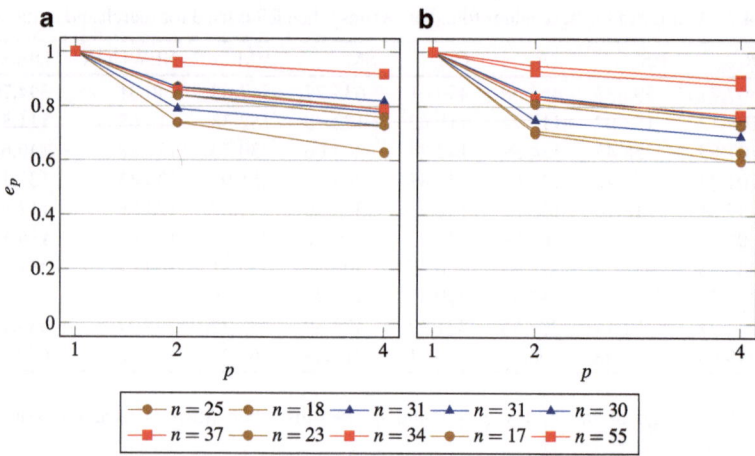

Fig. 4.8 Efficiency of parallel DISIMPL versions solving pile placement in grillage-type foundations: (a) DISIMPL-V algorithm, (b) DISIMPL-C algorithm

Appendix A
Description of Test Problems

Test Problems with Bound Constraints

- **Test Problem 1 [49]**

  - Number of variables: $n = 2$
  - Objective function:

  $$f(\mathbf{x}) = -4x_1x_2 \sin(4\pi x_2)$$

  - Feasible region: $\mathbb{D} = [0, 1]^2$
  - Accuracy: $\varepsilon = 0.355$
  - Global minimum: $f^* = -2.51997258$
  - Global minimizer: $\mathbf{x}^* = (1.00000000, 0.63492204)$
  - Derivatives:

  $$f_1'(\mathbf{x}) = -4x_2 \sin 4\pi x_2$$
  $$f_2'(\mathbf{x}) = -4x_1 \sin 4\pi x_2 - 16\pi x_1 x_2 \cos 4\pi x_2$$

- **Test Problem 2 [49]**

  - Number of variables: $n = 2$
  - Objective function:

  $$f(\mathbf{x}) = -\sin(2x_1 + 1) - 2\sin(3x_2 + 2)$$

  - Feasible region: $\mathbb{D} = [0, 1]^2$
  - Accuracy: $s = 0.0446$
  - Global minimum: $f^* = -2.81859485$
  - Global minimizer: $\mathbf{x}^* = (0.28539815, 0.00000000)$
  - Derivatives:

R. Paulavičius and J. Žilinskas, *Simplicial Global Optimization*,
SpringerBriefs in Optimization, DOI 10.1007/978-1-4614-9093-7,
© Remigijus Paulavičius, Julius Žilinskas 2014

$$f_1'(\mathbf{x}) = -2\cos(2x_1 + 1)$$
$$f_2'(\mathbf{x}) = -6\cos(3x_2 + 2)$$

- **Test Problem 3 (Branin) [23, 49]**

  - Number of variables: $n = 2$
  - Objective function:

$$f(\mathbf{x}) = \left(x_2 - \frac{5.1}{4\pi^2}x_1^2 + \frac{5}{\pi}x_1 - 6\right)^2 + 10\left(1 - \frac{1}{8\pi}\right)\cos(x_1) + 10$$

  - Feasible region: $\mathbb{D} = [-5, 10] \times [0, 15]$
  - Accuracy: $\varepsilon = 11.9$
  - Global minimum: $f^* = 0.397887$
  - Global minimizers: $\mathbf{x}^* = (−Š\pi, 12.275), (\pi, 2.275), (9.42478, 2.475)$
  - Derivatives:

$$f_1'(\mathbf{x}) = 2\left(\frac{5}{\pi} - \frac{2.55}{\pi^2}x_1\right)\left(-\frac{1.275}{\pi^2}x_1^2 + \frac{5}{\pi}x_1 + x_2 - 6\right)$$
$$+ 10(\sin x_1)\left(\frac{1}{8\pi} - 1\right)$$
$$f_2'(\mathbf{x}) = -\frac{2.55}{\pi^2}x_1^2 + \frac{10}{\pi}x_1 + 2x_2 - 12$$

- **Test Problem 4 [49]**

  - Number of variables: $n = 2$
  - Objective function:

$$f(\mathbf{x}) = \max\left(\sqrt{3}x_1 + x_2, -2x_2, x_2 - \sqrt{3}x_1\right)$$

  - Feasible region: $\mathbb{D} = [-1, 1]^2$
  - Accuracy: $\varepsilon = 0.0141$
  - Global minimum: $f^* = 0.00000000$
  - Global minimizer: $\mathbf{x}^* = (0.00000000, 0.00000000)$

- **Test Problem 5 [49]**

  - Number of variables: $n = 2$
  - Objective function:

$$f(\mathbf{x}) = -e^{-x_1^2}\sin x_1 + |x_2|$$

- Feasible region: $\mathbb{D} = [0, 10]^2$
- Accuracy: $\varepsilon = 0.1$
- Global minimum: $f^* = -0.39665295$
- Global minimizer: $\mathbf{x}^* = (0.65327118, 0.00000000)$
- Derivatives:

$$f_1'(\mathbf{x}) = -e^{-x_1^2} \cos x_1 + 2x_1 e^{-x_1^2} \sin x_1$$

$$f_2'(\mathbf{x}) = \operatorname{signum}(x_2)$$

• **Test Problem 6 [49]**

- Number of variables: $n = 2$
- Objective function:

$$f(\mathbf{x}) = 2x_1^2 - 1.05x_1^4 + x_2^2 - x_1 x_2 + \frac{1}{6}x_2^2$$

- Feasible region: $\mathbb{D} = [-2, 4]^2$
- Accuracy: $\varepsilon = 44.9$
- Global minimum: $f^* = -239.696629$
- Global minimizer: $\mathbf{x}^* = (4.00000000, -1.11976261)$
- Derivatives:

$$f_1'(\mathbf{x}) = -4.2x_1^3 + 4x_1 - x_2$$

$$f_2'(\mathbf{x}) = -x_1 + \frac{7}{3}x_2$$

• **Test Problem 7 [49]**

- Number of variables: $n = 2$
- Objective function:

$$f(\mathbf{x}) = \sum_{i=1}^{n-1} \left(100 \left(x_{i+1} - x_i^2 \right)^2 + (x_i - 1)^2 \right)$$

- Feasible region: $\mathbb{D} = [-3, 3] \times [-1.5, 4.5]$
- Accuracy: $\varepsilon = 542.0$
- Global minimum: $f^* = 0$
- Global minimizer: $\mathbf{x}^* = (1, 1)$
- Derivatives:

$$f_1'(\mathbf{x}) = -400x_1 \left(x_2 - x_1^2 \right) + 2x_1 - 2$$

$$f_2'(\mathbf{x}) = -200x_1^2 + 200x_2$$

- **Test Problem 8 [49]**

  - Number of variables: $n = 2$
  - Objective function:

  $$f(\mathbf{x}) = (x_1 - 2x_2 - 7)^2 + (2x_1 + x_2 - 5)^2$$

  - Feasible region: $\mathbb{D} = [-2.5, 3.5] \times [-1.5, 4.5]$
  - Accuracy: $\varepsilon = 3.66$
  - Global minimum: $f^* = 0.450000$
  - Global minimizer: $\mathbf{x}^* = (3.40000000, -1.49999999)$
  - Derivatives:

  $$f_1'(\mathbf{x}) = -34 + 10x_1$$
  $$f_2'(\mathbf{x}) = 10x_2 + 18$$

- **Test Problem 9 (Goldstein-Price) [23, 49]**

  - Number of variables: $n = 2$
  - Objective function:

  $$f(\mathbf{x}) = \left(1 + (x_1 + x_2 + 1)^2 \left(19 - 14x_1 + 3x_1^2 - 14x_2 + 6x_1x_2 + 3x_2^2\right)\right)$$
  $$\times \left(30 + (2x_1 - 3x_2)^2 \left(18 - 32x_1 + 12x_1^2 + 48x_2 - 36x_1x_2 + 27x_2^2\right)\right)$$

  - Feasible region: $\mathbb{D} = [-2, 2]^2$
  - Accuracy: $\varepsilon = 62900$
  - Global minimum: $f^* = 3.$
  - Global minimizer: $\mathbf{x}^* = (0, -1)$
  - Derivatives:

  $$
  \begin{aligned}
  f_1'(\mathbf{x}) = {} & 1152x_1^7 - 2016x_1^6x_2 - 5376x_1^6 - 3888x_1^5x_2^2 + 8064x_1^5x_2 \\
  & + 5712x_1^5 + 6120x_1^4x_2^3 + 12960x_1^4x_2^2 - 840x_1^4x_2 + 6720x_1^4 \\
  & + 5220x_1^3x_2^4 - 16320x_1^3x_2^3 - 21480x_1^3x_2^2 - 30720x_1^3x_2 - 9816x_1^3 \\
  & - 5508x_1^2x_2^5 - 10440x_1^2x_2^4 + 3720x_1^2x_2^3 + 29520x_1^2x_2^2 + 17352x_1^2x_2 \\
  & - 3216x_1^2 - 2916x_1x_2^6 + 7344x_1x_2^5 + 17460x_1x_2^4 + 10080x_1x_2^3 \\
  & + 15552x_1x_2^2 + 14688x_1x_2 + 2520x_1 + 972x_2^7 + 1944x_2^6 \\
  & - 1188x_2^5 - 11880x_2^4 - 23616x_2^3 - 19296x_2^2 - 4680x_2 + 720 \\
  f_2'(\mathbf{x}) = {} & -288x_1^7 - 1296x_1^6x_2 + 1344x_1^6 + 3672x_1^5x_2^2 + 5184x_1^5x_2 - 168x_1^5 \\
  & + 5220x_1^4x_2^3 - 12240x_1^4x_2^2 - 10740x_1^4x_2 - 7680x_1^4 - 9180x_1^3x_2^4
  \end{aligned}
  $$

$$- 13920x_1^3x_2^3 + 3720x_1^3x_2^2 + 19680x_1^3x_2 + 5784x_1^3 - 8748x_1^2x_2^5$$
$$+ 18360x_1^2x_2^4 + 34920x_1^2x_2^3 + 15120x_1^2x_2^2 + 15552x_1^2x_2$$
$$+ 7344x_1^2 + 6804x_1x_2^6 + 11664x_1x_2^5 - 5940x_1x_2^4 - 47520x_1x_2^3$$
$$- 70848x_1x_2^2 - 38592x_1x_2 - 4680x_1 + 5832x_2^7 - 4536x_2^6$$
$$- 26568x_2^5 + 9720x_2^4 + 57384x_2^3 + 36864x_2^2 + 6120x_2 + 720$$

- **Test Problem 10 [49]**

  - Number of variables: $n = 2$
  - Objective function:

  $$f(\mathbf{x}) = \sin(x_1 + x_2) + (x_1 - x_2)^2 - 1.5x_1 + 2.5x_2 + 1$$

  - Feasible region: $\mathbb{D} = [-1.5, 4] \times [-3, 3]$
  - Accuracy: $\varepsilon = 0.691$
  - Global minimum: $f^* = -1.91322295$
  - Global minimizer: $\mathbf{x}^* = (-0.54719386, -1.54720091)$
  - Derivatives:

  $$f_1'(\mathbf{x}) = -2x_2 + 2x_1 + \cos(x_1 - x_2) - 1.5$$
  $$f_2'(\mathbf{x}) = -2x_1 + 2x_2 + \cos(x_1 - x_2) + 2.5$$

- **Test Problem 11 [49]**

  - Number of variables: $n = 2$
  - Objective function:

  $$f(\mathbf{x}) = (x_1 - 2)^2 + (x_2 - 1)^2 + \frac{0.04}{-\frac{x_1^2}{4} - x_2^2 + 1} + \frac{(x_1 - 2x_2 + 1)^2}{0.2}$$

  - Feasible region: $\mathbb{D} = [1, 2]^2$
  - Accuracy: $\varepsilon = 0.335$
  - Global minimum: $f^* = 0.16904267$
  - Global minimizer: $\mathbf{x}^* = (1.79541003, 1.37786415)$
  - Derivatives:

  $$f_1'(\mathbf{x}) = -20x_2 + 12x_1 + 0.02\frac{x_1}{\left(\frac{1}{4}x_1^2 + x_2^2 - 1\right)^2} + 6$$

  $$f_2'(\mathbf{x}) = -20x_1 + 42x_2 + 0.08\frac{x_2}{\left(\frac{1}{4}x_1^2 + x_2^2 - 1\right)^2} - 22$$

- **Test Problem 12 [49]**

  - Number of variables: $n = 2$
  - Objective function:

$$f(\mathbf{x}) = 0.1\left(12 + x_1^2 + \frac{(1 + x_2^2)}{x_1^2} + \frac{x_1^2 x_2^2 + 100}{x_1^4 x_2^4}\right)$$

  - Feasible region: $\mathbb{D} = [1, 3]^2$
  - Accuracy: $\varepsilon = 0.804$
  - Global minimum: $f^* = 1.74$
  - Global minimizer: $\mathbf{x}^* = (1.74333181, 2.02987349)$
  - Derivatives:

$$f_1'(\mathbf{x}) = -\frac{0.2}{x_1^3}\left(x_2^2 + 1\right) + \frac{0.2}{x_1^3 x_2^2} - \frac{0.4}{x_1^5 x_2^4}\left(x_1^2 x_2^2 + 100\right) + 0.1$$

$$f_2'(\mathbf{x}) = -\frac{0.4}{x_1^4 x_2^5}\left(x_1^2 x_2^2 + 100\right) + \frac{0.2}{x_1^2 x_2^3} + \frac{0.2}{x_1^2} x_2$$

- **Test Problem 13 [49]**

  - Number of variables: $n = 2$
  - Objective function:

$$f(\mathbf{x}) = \frac{1}{2}\left(x_1^2 + x_2^2\right) - \cos\left(10 \ln\left(2x_1\right)\right) \cos\left(10 \ln\left(3x_2\right)\right) + 1$$

  - Feasible region: $\mathbb{D} = [0.01, 1]^2$
  - Accuracy: $\varepsilon = 6.92$
  - Global minimum: $f^* = -0.0001$
  - Global minimizer: $\mathbf{x}^* = (0.01152703, 0.01440453)$
  - Derivatives:

$$f_1'(\mathbf{x}) = x_1 + \frac{10}{x_1} \cos\left(10 \ln 3x_2\right) \sin\left(10 \ln 2x_1\right)$$

$$f_2'(\mathbf{x}) = x_2 + \frac{10}{x_2} \cos\left(10 \ln 2x_1\right) \sin\left(10 \ln 3x_2\right)$$

- **Test Problem 14 [49]**

  - Number of variables: $n = 3$
  - Objective function:

$$f(\mathbf{x}) = 100\left(x_3 - \left(\frac{x_1 + x_2}{2}\right)^2\right)^2 + (1 - x_1)^2 + (1 - x_2)^2$$

- Feasible region: $\mathbb{D} = [0, 1]^3$
- Accuracy: $\varepsilon = 2.12$
- Global minimum: $f^* = 0$
- Global minimizer: $\mathbf{x}^* = (1, 1, 1)$
- Derivatives:

$$f_1'(\mathbf{x}) = -200\left(x_3 - \left(\frac{1}{2}x_1 + \frac{1}{2}x_2\right)^2\right)\left(\frac{1}{2}x_1 + \frac{1}{2}x_2\right) + 2x_1 - 2$$

$$f_2'(\mathbf{x}) = -200\left(x_3 - \left(\frac{1}{2}x_1 + \frac{1}{2}x_2\right)^2\right)\left(\frac{1}{2}x_1 + \frac{1}{2}x_2\right) + 2x_2 - 2$$

$$f_3'(\mathbf{x}) = -200\left(\frac{1}{2}x_1 + \frac{1}{2}x_2\right)^2 + 200x_3$$

- **Test Problem 15 (Hartman-3) [23, 49]**

  - Number of variables: $n = 3$
  - Objective function:

$$f(\mathbf{x}) = -\sum_{i=1}^{4} c_i \exp\left(-\sum_{j=1}^{3} \alpha_{ij}\left(x_j - p_{ij}\right)^2\right)$$

  - Feasible region: $\mathbb{D} = [0, 1]^3$
  - Accuracy: $\varepsilon = 0.369$
  - Global minimum: $f^* = -3.8627814$
  - Global minimizer: $\mathbf{x}^* = (0.114540, 0.555784, 0.852538)$
  - Derivatives:

$$f_1'(\mathbf{x}) = \alpha_{11}c_1 \exp\left(-\alpha_{11}(x_1 - p_{11})^2 - \alpha_{12}(x_2 - p_{12})^2 - \alpha_{13}(x_3 - p_{13})^2\right)$$

$$\times (2x_1 - 2p_{11}) + \alpha_{21}c_2$$

$$\times \exp\left(-\alpha_{21}(x_1 - p_{21})^2 - \alpha_{22}(x_2 - p_{22})^2 - \alpha_{23}(x_3 - p_{23})^2\right)$$

$$\times (2x_1 - 2p_{21}) + \alpha_{31}c_3$$

$$\times \exp\left(-\alpha_{31}(x_1 - p_{31})^2 - \alpha_{32}(x_2 - p_{32})^2 - \alpha_{33}(x_3 - p_{33})^2\right)$$

$$\times (2x_1 - 2p_{31}) + \alpha_{41}c_4$$

$$\times \exp\left(-\alpha_{41}(x_1 - p_{41})^2 - \alpha_{42}(x_2 - p_{42})^2 - \alpha_{43}(x_3 - p_{43})^2\right)$$

$$\times (2x_1 - 2p_{41})$$

$$f_2'(\mathbf{x}) = \alpha_{12}c_1 \exp\left(-\alpha_{11}(x_1 - p_{11})^2 - \alpha_{12}(x_2 - p_{12})^2 - \alpha_{13}(x_3 - p_{13})^2\right)$$

$$\times (2x_2 - 2p_{12}) + \alpha_{22}c_2$$

$$\times \exp\left(-\alpha_{21}(x_1 - p_{21})^2 - \alpha_{22}(x_2 - p_{22})^2 - \alpha_{23}(x_3 - p_{23})^2\right)$$

$$\times (2x_2 - 2p_{22}) + \alpha_{32}c_3$$

$$\times \exp\left(-\alpha_{31}(x_1 - p_{31})^2 - \alpha_{32}(x_2 - p_{32})^2 - \alpha_{33}(x_3 - p_{33})^2\right)$$

$$\times (2x_2 - 2p_{32}) + \alpha_{42}c_4$$

$$\times \exp\left(-\alpha_{41}(x_1 - p_{41})^2 - \alpha_{42}(x_2 - p_{42})^2 - \alpha_{43}(x_3 - p_{43})^2\right)$$

$$\times (2x_2 - 2p_{42})$$

$$f_3'(\mathbf{x}) = \alpha_{13}c_1 \exp\left(-\alpha_{11}(x_1 - p_{11})^2 - \alpha_{12}(x_2 - p_{12})^2 - \alpha_{13}(x_3 - p_{13})^2\right)$$

$$\times (2x_3 - 2p_{13}) + \alpha_{23}c_2$$

$$\times \exp\left(-\alpha_{21}(x_1 - p_{21})^2 - \alpha_{22}(x_2 - p_{22})^2 - \alpha_{23}(x_3 - p_{23})^2\right)$$

$$\times (2x_3 - 2p_{23}) + \alpha_{33}c_3$$

$$\times \exp\left(-\alpha_{31}(x_1 - p_{31})^2 - \alpha_{32}(x_2 - p_{32})^2 - \alpha_{33}(x_3 - p_{33})^2\right)$$

$$\times (2x_3 - 2p_{33}) + \alpha_{43}c_4$$

$$\times \exp\left(-\alpha_{41}(x_1 - p_{41})^2 - \alpha_{42}(x_2 - p_{42})^2 - \alpha_{43}(x_3 - p_{43})^2\right)$$

$$\times (2x_3 - 2p_{43})$$

where

i	c_i	α_{i1}	α_{i2}	α_{i3}	p_{i1}	p_{i2}	p_{i3}
1	1.0	3.0	10	30	0.3689	0.1170	0.2673
2	1.2	0.1	10	35	0.4699	0.4387	0.7470
3	3.0	3.0	10	30	0.1091	0.8732	0.5547
4	3.2	1.0	10	35	0.0382	0.5743	0.8828

- **Test Problem 16 [49]**

  - Number of variables: $n = 3$
  - Objective function:

$$f(\mathbf{x}) = \frac{1}{3}(x_1^2 + x_2^2 + x_3^2) - \cos(10\ln(2x_1))\cos(10\ln(3x_2)) \times \cos(10\ln(4x_3)) + 1$$

- Feasible region: $\mathbb{D} = [0.01, 1]^3$
- Accuracy: $\varepsilon = 8.333$
- Global minimum: $f^* = 0$
- Global minimizer: $\mathbf{x}^* = (0.040555, 0.069074, 0.052778)$
- Derivatives:

$$f_1'(\mathbf{x}) = \frac{2}{3}x_1 + \frac{10}{x_1}\cos(10\ln 3x_2)\cos(10\ln 4x_3)\sin(10\ln 2x_1)$$

$$f_2'(\mathbf{x}) = \frac{2}{3}x_2 + \frac{10}{x_2}\cos(10\ln 2x_1)\cos(10\ln 4x_3)\sin(10\ln 3x_2)$$

$$f_3'(\mathbf{x}) = \frac{2}{3}x_3 + \frac{10}{x_3}\cos(10\ln 2x_1)\cos(10\ln 3x_2)\sin(10\ln 4x_3)$$

- **Test Problem 17 [49]**

  - Number of variables: $n = 3$
  - Objective function:

$$f(\mathbf{x}) = -\sin x_1 \sin x_1 x_2 \sin x_1 x_2 x_3$$

  - Feasible region: $\mathbb{D} = [0, 4]^3$
  - Accuracy: $\varepsilon = 0.672$
  - Global minimum: $f^* = -0.9$
  - Global minimizer: $\mathbf{x}^* = (1.572016, 2.998628, 2.998628)$
  - Derivatives:

$$f_1'(\mathbf{x}) = -\sin x_1 x_2 \cos x_1 \sin x_1 x_2 x_3 - x_2 \cos x_1 x_2 \sin x_1 \sin x_1 x_2 x_3$$
$$- x_2 x_3 \sin x_1 x_2 \sin x_1 \cos x_1 x_2 x_3$$
$$f_2'(\mathbf{x}) = -x_1 \cos x_1 x_2 \sin x_1 \sin x_1 x_2 x_3 - x_1 x_3 \sin x_1 x_2 \sin x_1 \cos x_1 x_2 x_3$$
$$f_3'(\mathbf{x}) = -x_1 x_2 \sin x_1 x_2 \sin x_1 \cos x_1 x_2 x_3$$

- **Test Problem 18 [49]**

  - Number of variables: $n = 3$
  - Objective function:

$$f(\mathbf{x}) = -\left(x_1^2 - 2x_2^2 + x_3^2\right)\sin x_1 \sin x_2 \sin x_3$$

  - Feasible region: $\mathbb{D} = [-1, 1]^3$
  - Accuracy: $\varepsilon = 0.0506$
  - Global minimum: $f^* = -0.51637406$
  - Global minimizer: $\mathbf{x}^* = (-1.000000, -0.555968, -1.000000)$
  - Derivatives:

$$f_1'(\mathbf{x}) = - (\cos x_1 \sin x_2 \sin x_3) \left(x_1^2 - 2x_2^2 + x_3^2\right) - 2x_1 \sin x_1 \sin x_2 \sin x_3$$
$$f_2'(\mathbf{x}) = - (\cos x_2 \sin x_1 \sin x_3) \left(x_1^2 - 2x_2^2 + x_3^2\right) + 4x_2 \sin x_1 \sin x_2 \sin x_3$$
$$f_3'(\mathbf{x}) = - (\cos x_3 \sin x_1 \sin x_2) \left(x_1^2 - 2x_2^2 + x_3^2\right) - 2x_3 \sin x_1 \sin x_2 \sin x_3$$

- **Test Problem 19 [49]**

  - Number of variables: $n = 3$
  - Objective function:

$$f(\mathbf{x}) = - (x_1 - 1)(x_1 + 2)(x_2 + 1)(x_2 - 2) x_3^2$$

  - Feasible region: $\mathbb{D} = [-2, 2]^3$
  - Accuracy: $\varepsilon = 4.51$
  - Global minimum: $f^* = 35.99999997$
  - Global minimizer: $\mathbf{x}^* = (-0.500000, -2.000000, -2.000000)$
  - Derivatives:

$$f_1'(\mathbf{x}) = -x_3^2 (x_1 - 1)(x_2 + 1)(x_2 - 2) - x_3^2 (x_1 + 2)(x_2 + 1)(x_2 - 2)$$
$$f_2'(\mathbf{x}) = -x_3^2 (x_1 - 1)(x_1 + 2)(x_2 + 1) - x_3^2 (x_1 - 1)(x_1 + 2)(x_2 - 2)$$
$$f_3'(\mathbf{x}) = -2x_3 (x_1 - 1)(x_1 + 2)(x_2 + 1)(x_2 - 2)$$

- **Test Problem 20 (Rosenbrock) [82]**

  - Number of variables: $n = 3$
  - Objective function:

$$f(\mathbf{x}) = \sum_{i=1}^{n-1} \left(100 \left(x_{i+1} - x_i^2\right)^2 + (x_i - 1)^2\right)$$

  - Feasible region: $\mathbb{D} = [-3, 3]^3$
  - Accuracy: $\varepsilon = 2500$
  - Global minimum: $f^* = 0$
  - Global minimizer: $\mathbf{x}^* = (1, 1, 1)$
  - Derivatives:

$$f_1'(\mathbf{x}) = -400 \left(x_2 - x_1^2\right) x_1 + 2 (x_1 - 1)$$
$$f_2'(\mathbf{x}) = -400 \left(x_{i+1} - x_i^2\right) x_i + 2 (x_{i+1} - 1) + 200 \left(x_i - x_{i-1}^2\right)$$
$$f_3'(\mathbf{x}) = 200 \left(x_n - x_{n-1}^2\right)$$

- **Test Problem 21 (Levy 15) [60]**

  - Number of variables: $n = 4$

- Objective function:

$$f(\mathbf{x}) = \sin^2 3\pi x_1 + \sum_{i=1}^{n-1} (x_i - 1)^2 \left(1 + \sin^2 3\pi x_{i+1}\right)$$

$$+ (x_n - 1)^2 \times \left(1 + \sin^2 2\pi x_n\right)$$

- Feasible region: $\mathbb{D} = [-10, 10]^4$
- Accuracy: $\varepsilon = L_2$
- Global minimum: $f^* = 0$
- Global minimizer: $\mathbf{x}^* = (1, 1, 1, 1)$
- Derivatives:

$$f_1'(\mathbf{x}) = 6\sin(3\pi x_1)\cos(3\pi x_1)\,\pi + 2(x_1 - 1)\left(1 + \sin^2(3\pi x_2)\right)$$

$$f_i'(\mathbf{x}) = 6(x_{i-1} - 1)^2 \sin(3\pi x_i)\cos(3\pi x_i)\,\pi + 2(x_i - 1)$$

$$\times \left(1 + \sin^2(3\pi x_{i+1})\right)$$

$$f_n'(\mathbf{x}) = 6(x_{n-1} - 1)^2 \sin(3\pi x_n)\cos(3\pi x_n)\,\pi + 1 + \sin^2(2\pi x_n)$$

$$+ 4(x_n - 1)\sin(2\pi x_n)\cos(2\pi x_n)$$

• **Test Problem 22 (Rosenbrock) [82]**

- Number of variables: $n = 4$
- Objective function:

$$f(\mathbf{x}) = \sum_{i=1}^{n-1} \left(100\left(x_{i+1} - x_i^2\right)^2 + (x_i - 1)^2\right)$$

- Feasible region: $\mathbb{D} = [-4, 4]^4$
- Accuracy: $\varepsilon = L_2$
- Global minimum: $f^* = 0$
- Global minimizer: $\mathbf{x}^* = (1, 1, 1, 1)$
- Derivatives:

$$f_1'(\mathbf{x}) = -400\left(x_2 - x_1^2\right)x_1 + 2(x_1 - 1)$$

$$f_i'(\mathbf{x}) = -400\left(x_{i+1} - x_i^2\right)x_i + 2(x_{i+1} - 1) + 200\left(x_i - x_{i-1}^2\right)$$

$$f_n'(\mathbf{x}) = 200\left(x_n - x_{n-1}^2\right)$$

• **Test Problem 23 (Shekel-5) [23, 60]**

- Number of variables: $n = 4$

– Objective function:

$$f(\mathbf{x}) = -\sum_{i=1}^{5} \frac{1}{(x - a_i)(x - a_i)^T + c_i}$$

– Feasible region: $\mathbb{D} = [0, 10]^4$
– Accuracy: $\varepsilon = L_2$
– Global minimum: $f^* = -10.1532$
– Global minimizer: $\mathbf{x}^* = (4.00004, 4.00013, 4.00004, 4.00013)$
– Derivatives:

$$f_j'(\mathbf{x}) = -\sum_{i=1}^{5} \frac{2(x_j - a_{i,j})}{\left((x - a_i)(x - a_i)^T + c_i\right)^2}$$

where

i	a_{i1}	a_{i2}	a_{i3}	a_{i4}	c_i
1	4.0	4.0	4.0	4.0	0.1
2	1.0	1.0	1.0	1.0	0.2
3	8.0	8.0	8.0	8.0	0.2
4	6.0	6.0	6.0	6.0	0.4
5	3.0	7.0	3.0	7.0	0.4

- **Test Problem 24 (Shekel-7) [23, 60]**

  – Number of variables: $n = 4$
  – Objective function:

$$f(\mathbf{x}) = -\sum_{i=1}^{7} \frac{1}{(x - a_i)(x - a_i)^T + c_i}$$

  – Feasible region: $\mathbb{D} = [0, 10]^4$
  – Accuracy: $\varepsilon = L_2$
  – Global minimum: $f^* = -10.4029$
  – Global minimizer: $\mathbf{x}^* = (4.00057, 4.00069, 3.99949, 3.99961)$
  – Derivatives:

$$f_j'(\mathbf{x}) = -\sum_{i=1}^{7} \frac{2(x_j - a_{i,j})}{\left((x - a_i)(x - a_i)^T + c_i\right)^2}$$

  where

i	a_{i1}	a_{i2}	a_{i3}	a_{i4}	c_i
1	4.0	4.0	4.0	4.0	0.1
2	1.0	1.0	1.0	1.0	0.2
3	8.0	8.0	8.0	8.0	0.2
4	6.0	6.0	6.0	6.0	0.4
5	3.0	7.0	3.0	7.0	0.4
6	2.0	9.0	2.0	9.0	0.6
7	5.0	5.0	3.0	3.0	0.6

- **Test Problem 25 (Shekel-10) [23, 60]**

  - Number of variables: $n = 4$
  - Objective function:

$$f(\mathbf{x}) = -\sum_{i=1}^{10} \frac{1}{(x - a_i)(x - a_i)^T + c_i}$$

  - Feasible region: $\mathbb{D} = [0, 10]^4$
  - Accuracy: $\varepsilon = L_2$
  - Global minimum: $f^* = -10.5364$
  - Global minimizer: $\mathbf{x}^* = (4.00075, 4.00059, 3.99966, 3.99951)$
  - Derivatives:

$$f'_j(\mathbf{x}) = -\sum_{i=1}^{10} \frac{2(x_j - a_{i,j})}{\left((x - a_i)(x - a_i)^T + c_i\right)^2}$$

  where

i	a_{i1}	a_{i2}	a_{i3}	a_{i4}	c_i
1	4.0	4.0	4.0	4.0	0.1
2	1.0	1.0	1.0	1.0	0.2
3	8.0	8.0	8.0	8.0	0.2
4	6.0	6.0	6.0	6.0	0.4
5	3.0	7.0	3.0	7.0	0.4
6	2.0	9.0	2.0	9.0	0.6
7	5.0	5.0	3.0	3.0	0.6
8	8.0	1.0	8.0	1.0	0.7
9	6.0	2.0	6.0	2.0	0.5
10	7.0	3.6	7.0	3.6	0.5

- **Test Problem 26 (Schwefel 1.2) [60]**

    - Number of variables: $n = 4$
    - Objective function:

$$f(\mathbf{x}) = \sum_{i=1}^{4} \left(\sum_{j=1}^{i} x_j \right)^2$$

    - Feasible region: $\mathbb{D} = [-5, 10]^4$
    - Accuracy: $\varepsilon = L_2$
    - Global minimum: $f^* = 0$
    - Global minimizer: $\mathbf{x}^* = (1, 1, 1, 1)$
    - Derivatives:

$$f_1'(\mathbf{x}) = 8x_1 + 6x_2 + 4x_3 + 2x_4$$
$$f_2'(\mathbf{x}) = 6x_1 + 6x_2 + 4x_3 + 2x_4$$
$$f_3'(\mathbf{x}) = 4x_1 + 4x_2 + 4x_3 + 2x_4$$
$$f_4'(\mathbf{x}) = 2x_1 + 2x_2 + 2x_3 + 2x_4$$

- **Test Problem 27 (Powel) [60]**

    - Number of variables: $n = 4$
    - Objective function:

$$f(\mathbf{x}) = (x_1 + 10x_2)^2 + 5(x_3 - x_4)^2 + (x_2 - 2x_3)^4 + 10(x_1 - x_4)^4$$

    - Feasible region: $\mathbb{D} = [-4, 5]^4$
    - Accuracy: $\varepsilon = L_2$
    - Global minimum: $f^* = 0$
    - Global minimizer: $\mathbf{x}^* = (0, 0, 0, 0)$
    - Derivatives:

$$f_1'(\mathbf{x}) = 2x_1 + 20x_2 + 40(x_1 - x_4)^3$$
$$f_2'(\mathbf{x}) = 20x_1 + 200x_2 + 4(x_2 - 2x_3)^3$$
$$f_3'(\mathbf{x}) = -10x_4 + 10x_3 - 8(x_2 - 2x_3)^3$$
$$f_4'(\mathbf{x}) = -10x_3 + 10x_4 - 40(x_1 + x_4)^3$$

- **Test Problem 28 (Levy 9) [60]**

    - Number of variables: $n = 4$
    - Objective function:

$$f(\mathbf{x}) = \sin^2 3\pi y_1 + \sum_{i=1}^{n-1} (y_i - 1)^2 \left(1 + 10\sin^2 \pi y_{i+1}\right) + (y_n - 1)^2,$$

$$y_i = 1 + (x_i - 1)/4$$

- Feasible region: $\mathbb{D} = [-10, 10]^4$
- Accuracy: $\varepsilon = L_2$
- Global minimum: $f^* = 0$
- Global minimizer: $\mathbf{x}^* = (1, 1, 1, 1)$
- Derivatives:

$$f_1'(\mathbf{x}) = \left(10\sin^2 \pi \left(\frac{1}{4}x_2 + \frac{3}{4}\right) + 1\right)(2x_1 - 2)$$
$$+ \frac{3}{2}\pi \cos 3\pi \left(\frac{1}{4}x_1 + \frac{3}{4}\right) \sin 3\pi \left(\frac{1}{4}x_1 + \frac{3}{4}\right)$$

$$f_2'(\mathbf{x}) = \left(10\sin^2 \pi \left(\frac{1}{4}x_4 + \frac{3}{4}\right) + 1\right)\left(\frac{1}{8}x_3 - \frac{1}{8}\right)$$
$$+ 5\pi \left(\cos \pi \left(\frac{1}{4}x_3 + \frac{3}{4}\right) \sin \pi \left(\frac{1}{4}x_3 + \frac{3}{4}\right)\right)\left(\frac{1}{4}x_2 - \frac{1}{4}\right)^2$$

$$f_3'(\mathbf{x}) = \frac{1}{8} + 5\pi \left(\cos \pi \left(\frac{1}{4}x_4 + \frac{3}{4}\right) \sin \pi \left(\frac{1}{4}x_4 + \frac{3}{4}\right)\right)\left(\frac{1}{4}x_3 - \frac{1}{4}\right)^2 + \frac{1}{8}x_4$$

- **Test Problem 29 (Levy 16) [60]**

  - Number of variables: $n = 5$
  - Objective function:

$$f(\mathbf{x}) = \sin^2 3\pi x_1 + \sum_{i=1}^{n-1} (x_i - 1)^2 \left(1 + \sin^2 3\pi x_{i+1}\right)$$

$$+ (x_n - 1)^2 \left(1 + \sin^2 2\pi x_n\right)$$

  - Feasible region: $\mathbb{D} = [-5, 5]^5$
  - Accuracy: $\varepsilon = 1.5L_2$
  - Global minimum: $f^* = 0$
  - Global minimizer: $\mathbf{x}^* = (1, 1, 1, 1, 1)$
  - Derivatives:

$$f_1'(\mathbf{x}) = 6\sin(3\pi x_1)\cos(3\pi x_1)\,\pi + 2(x_1 - 1)\left(1 + \sin^2(3\pi x_2)\right)$$
$$f_i'(\mathbf{x}) = 6(x_{i-1} - 1)^2 \sin(3\pi x_i)\cos(3\pi x_i)\,\pi$$
$$+ 2(x_i - 1) \times \left(1 + \sin^2(3\pi x_{i+1})\right)$$

$$f_n'(\mathbf{x}) = 6\,(x_{n-1} - 1)^2 \sin\,(3\pi x_n) \cos\,(3\pi x_n)\,\pi + 1 + \sin^2\,(2\pi x_n)$$
$$\qquad\quad + 4\,(x_n - 1) \sin\,(2\pi x_n) \cos\,(2\pi x_n)$$

- **Test Problem 30 (Rosenbrock) [82]**

  - Number of variables: $n = 5$
  - Objective function:

$$f(\mathbf{x}) = \sum_{i=1}^{n-1} \left(100\,\left(x_{i+1} - x_i^2\right)^2 + (x_i - 1)^2\right)$$

  - Feasible region: $\mathbb{D} = [-5, 5]^5$
  - Accuracy: $\varepsilon = 1.5 L_2$
  - Global minimum: $f^* = 0$
  - Global minimizer: $\mathbf{x}^* = (1, 1, 1, 1, 1)$
  - Derivatives:

$$f_1'(\mathbf{x}) = -400\,\left(x_2 - x_1^2\right) x_1 + 2\,(x_1 - 1)$$
$$f_i'(\mathbf{x}) = -400\,\left(x_{i+1} - x_i^2\right) x_i + 2\,(x_{i+1} - 1) + 200\,\left(x_i - x_{i-1}^2\right)$$
$$f_n'(\mathbf{x}) = 200\,\left(x_n - x_{n-1}^2\right)$$

- **Test Problem 31 (Levy 10) [60]**

  - Number of variables: $n = 5$
  - Objective function:

$$f(\mathbf{x}) = \sin^2 3\pi y_1 + \sum_{i=1}^{n-1} (y_i - 1)^2 \left(1 + 10 \sin^2 \pi y_{i+1}\right) + (y_n - 1)^2,$$
$$y_i = 1 + (x_i - 1)/4$$

  - Feasible region: $\mathbb{D} = [-10, 10]^5$
  - Accuracy: $\varepsilon = 1.5 L_2$
  - Global minimum: $f^* = 0$
  - Global minimizer: $\mathbf{x}^* = (1, 1, 1, 1, 1)$
  - Derivatives:

$$f_1'(\mathbf{x}) = \left(10 \sin^2 \pi \left(\frac{1}{4}x_2 + \frac{3}{4}\right) + 1\right) (2x_1 - 2)$$
$$\qquad + \frac{3}{2}\pi \cos 3\pi \left(\frac{1}{4}x_1 + \frac{3}{4}\right) \sin 3\pi \left(\frac{1}{4}x_1 + \frac{3}{4}\right)$$

$$f_2'(\mathbf{x}) = \left(10\sin^2 \pi \left(\frac{1}{4}x_3 + \frac{3}{4}\right) + 1\right)\left(\frac{1}{8}x_2 - \frac{1}{8}\right)$$

$$+ 5\pi \left(\cos \pi \left(\frac{1}{4}x_2 + \frac{3}{4}\right)\sin \pi \left(\frac{1}{4}x_2 + \frac{3}{4}\right)\right)(x_1 - 1)^2$$

$$f_i'(\mathbf{x}) = \left(10\sin^2 \pi \left(\frac{1}{4}x_{i+1} + \frac{3}{4}\right) + 1\right)\left(\frac{1}{8}x_i - \frac{1}{8}\right)$$

$$+ 5\pi \left(\cos \pi \left(\frac{1}{4}x_i + \frac{3}{4}\right)\sin \pi \left(\frac{1}{4}x_i + \frac{3}{4}\right)\right)\left(\frac{1}{4}x_{i-1} - \frac{1}{4}\right)$$

$$f_n'(\mathbf{x}) = -\frac{1}{8} + 5\pi \left(\cos \pi \left(\frac{1}{4}x_n + \frac{3}{4}\right)\sin \pi \left(\frac{1}{4}x_n + \frac{3}{4}\right)\right)\left(\frac{1}{4}x_{n-1} - \frac{1}{4}\right)^2$$

$$+ \frac{1}{8}x_n$$

- **Test Problem 32 (Levy 17) [60]**

  - Number of variables: $n = 6$
  - Objective function:

  $$f(\mathbf{x}) = \sin^2 3\pi x_1 + \sum_{i=1}^{n-1} (x_i - 1)^2 \left(1 + \sin^2 3\pi x_{i+1}\right)$$

  $$+ (x_n - 1)^2 \left(1 + \sin^2 2\pi x_n\right)$$

  - Feasible region: $\mathbb{D} = [-5, 5]^6$
  - Accuracy: $\varepsilon = 4L_2$
  - Global minimum: $f^* = 0$
  - Global minimizer: $\mathbf{x}^* = (1, 1, 1, 1, 1, 1)$
  - Derivatives:

  $$f_1'(\mathbf{x}) = 6\sin(3\pi x_1)\cos(3\pi x_1)\pi + 2(x_1 - 1)\left(1 + \sin^2(3\pi x_2)\right)$$

  $$f_i'(\mathbf{x}) = 6(x_{i-1} - 1)^2\sin(3\pi x_i)\cos(3\pi x_i)\pi$$

  $$+2(x_i - 1) \times \left(1 + \sin^2(3\pi x_{i+1})\right)$$

  $$f_n'(\mathbf{x}) = 6(x_{n-1} - 1)^2\sin(3\pi x_n)\cos(3\pi x_n)\pi + 1 + \sin^2(2\pi x_n)$$

  $$+ 4(x_n - 1)\sin(2\pi x_n)\cos(2\pi x_n)$$

- **Test Problem 33 (Rosenbrock) [82]**

  - Number of variables: $n = 6$
  - Objective function:

  $$f(\mathbf{x}) = \sum_{i=1}^{n-1} \left(100\left(x_{i+1} - x_i^2\right)^2 + (x_i - 1)^2\right)$$

- Feasible region: $\mathbb{D} = [-6, 6]^6$
- Accuracy: $\varepsilon = 4L_2$
- Global minimum: $f^* = 0$
- Global minimizer: $\mathbf{x}^* = (1, 1, 1, 1, 1, 1)$
- Derivatives:

$$f_1'(\mathbf{x}) = -400\left(x_2 - x_1^2\right)x_1 + 2\left(x_1 - 1\right)$$

$$f_i'(\mathbf{x}) = -400\left(x_{i+1} - x_i^2\right)x_i + 2\left(x_{i+1} - 1\right) + 200\left(x_i - x_{i-1}^2\right)$$

$$f_n'(\mathbf{x}) = 200\left(x_n - x_{n-1}^2\right)$$

- **Test Problem Six-Hump Camelback [143]**

  - Number of variables: $n = 2$
  - Objective function:

$$f(\mathbf{x}) = \left(4 - 2.1x_1^2 + \frac{1}{3}x_1^4\right)x_1^2 + x_1x_2 + (-4 + 4x_2^2)x_2^2$$

  - Feasible region: $\mathbb{D} = [-3, 3] \times [-2, 2]$
  - Global minimum: $f^* = -1.031628$
  - Global minimizers: $\mathbf{x}^* = (0.089842, -0.712656), (-0.089842, 0.712656)$

- **Test Problem Shubert [143]**

  - Number of variables: $n = 2$
  - Objective function:

$$f(\mathbf{x}) = \left(\sum_{i=1}^{5} i \cos\left((i+1)x_1 + i\right)\right) \times \left(\sum_{i=1}^{5} i \cos\left((i+1)x_2 + i\right)\right)$$

  - Feasible region: $\mathbb{D} = [-10, 10]^2$
  - Global minimum: $f^* = -186.7309$
  - Global minimizers: 18 global minimizers

- **Test Problem Alolyan [46]**

  - Number of variables: $n = 2$
  - Objective function:

$$f(\mathbf{x}) = x_1x_2^2 + x_2x_1^2 - x_1^3 - x_2^3$$

  - Feasible region: $\mathbb{D} = [-1, 1]^2$
  - Global minimum: $f^* = âĹŠ1.18519$
  - Global minimizers: $\mathbf{x}^* = (-\frac{1}{3}, 1), (\frac{1}{3}, -1)$

- **Test Problem Easom [46]**

  - Number of variables: $n = 2$
  - Objective function:

  $$f(\mathbf{x}) = -\cos x_1 \cos x_2 \exp(-(x_1 - \pi)^2 - (x_2 - \pi)^2)$$

  - Feasible region: $\mathbb{D} = [-100, -100]$
  - Global minimum: $f^* = -1$
  - Global minimizer: $\mathbf{x}^* = (\pi, \pi)$

- **Test Problem Rastrigin [46]**

  - Number of variables: $n = 2$
  - Objective function:

  $$f(\mathbf{x}) = 20 + x_1^2 + x_2^2 - 10(\cos 2\pi x_1 + \cos 2\pi x_2)$$

  - Feasible region: $\mathbb{D} = [-5, 6]^2$
  - Global minimum: $f^* = 0$
  - Global minimizer: $\mathbf{x}^* = (0, 0)$

- **Test Problem Hartman-6 [23]**

  - Number of variables: $n = 6$
  - Objective function:

  $$f(\mathbf{x}) = -\sum_{i=1}^{4} c_i \exp\left(-\sum_{j=1}^{6} \alpha_{ij} \left(x_j - p_{ij}\right)^2\right)$$

  where

i	c_i	α_{i1}	α_{i2}	α_{i3}	α_{i4}	α_{i5}	α_{i6}
1	1.0	10.00	3.00	17.00	3.50	1.70	8.00
2	1.2	0.05	10.00	17.00	0.10	8.00	14.00
3	3.0	3.00	3.50	1.70	10.00	17.00	8.00
4	3.2	17.00	8.00	0.05	10.00	0.10	14.00

i	p_{i1}	p_{i2}	p_{i3}	p_{i4}	p_{i5}	p_{i6}
1	0.1312	0.1696	0.5569	0.0124	0.8283	0.5886
2	0.2329	0.4135	0.8307	0.3736	0.1004	0.9991
3	0.2348	0.1451	0.3522	0.2883	0.3047	0.6650
4	0.4047	0.8828	0.8732	0.5743	0.1091	0.0381

  - Feasible region: $\mathbb{D} = [0, 1]^6$
  - Accuracy: $\varepsilon = 0.369$

- Global minimum: $f^* = âĹŠ3.32237$
- Global minimizer:

$$\mathbf{x}^* = (0.20169, 0.15001, 0.47687, 0.27533, 0.31165, 0.65730)$$

Test Problems with Linear Constraints

- **Test Problem Horst 1 [55]**

  - Number of variables: $n = 2$
  - Objective function:

  $$f(\mathbf{x}) = -x_1^2 - 4x_2^2 + 4x_1x_2 + 2x_1 + 4x_2$$

  - Constraints:

  $$-4x_1 + 2x_2 \leq 1,$$
  $$x_1 + x_2 \leq 4,$$
  $$x_1 - 4x_2 \leq 1,$$
  $$0 \leq x_1 \leq 3,$$
  $$0 \leq x_2 \leq 2.$$

  - Global minimum: $f^* = -1.0625$
  - Global minimizer: $\mathbf{x}^* = (0.75, 2.0)$

- **Test Problem Horst 2 [55]**

  - Number of variables: $n = 2$
  - Objective function:

  $$f(\mathbf{x}) = -x_1^2 - x_2^{3/2}$$

  - Constraints:

  $$x_1 + 2x_2 \leq 4,$$
  $$x_1 - 2x_2 \leq 1,$$
  $$-x_1 + x_2 \leq 1,$$
  $$0 \leq x_1 \leq 2.5,$$
  $$0 \leq x_2 \leq 2.$$

- Global minimum: $f^* = -6.8995$
- Global minimizer: $\mathbf{x}^* = (2.5, 0.75)$

- **Test Problem Horst 3 [55]**

  - Number of variables: $n = 2$
  - Objective function:

$$f(\mathbf{x}) = -x_1^2 + \frac{4}{3}x_1 + \ln(1 + x_2) - \frac{4}{9}$$

  - Constraints:

$$-2x_1 + x_2 \leq 1,$$
$$x_1 + x_2 \leq \frac{3}{2},$$
$$x_1 + \frac{1}{10}x_2 \leq 1,$$
$$x_1, x_2 \geq 0.$$

  - Global minimum: $f^* = -\frac{4}{9}$
  - Global minimizer: $\mathbf{x}^* = (0.0, 0.0)$

- **Test Problem Horst 4 [55]**

  - Number of variables: $n = 3$
  - Objective function:

$$f(\mathbf{x}) = -|x_1 + \frac{1}{2}x_2 + \frac{2}{3}x_3|^{\frac{3}{2}}$$

  - Constraints:

$$x_1 + x_2 + 2x_3 \leq 6,$$
$$x_1 + \frac{1}{2}x_2 \leq 2,$$
$$-x_2 - 2x_3 \leq -1,$$
$$-x_1 \leq -\frac{1}{2},$$
$$x_1, x_2, x_3 \geq 0.$$

  - Global minimum: $f^* = -6.0858$
  - Global minimizer: $\mathbf{x}^* = (2.0, 0.0, 2.0)$

- **Test Problem Horst 5 [55]**

  - Number of variables: $n = 3$
  - Objective function:

$$f(\mathbf{x}) = -|x_1 + \frac{1}{2}x_2 + \frac{2}{3}x_3|^{\frac{3}{2}} - x_1^2$$

  - Constraints:

$$x_1 + x_2 + x_3 \leq 2,$$
$$x_1 + x_2 - \frac{1}{4}x_3 \leq 1,$$
$$-2x_1 - 2x_2 + x_3 \leq 1,$$
$$x_3 \leq 3,$$
$$x_1, x_2, x_3 \geq 0.$$

  - Global minimum: $f^* = -3.722$
  - Global minimizer: $\mathbf{x}^* = (1.2, 0.0, 0.8)$

- **Test Problem Horst 6 [55]**

  - Number of variables: $n = 3$
  - Objective function:

$$f(\mathbf{x}) = \mathbf{x}^T \mathbf{Q} \mathbf{x} + \mathbf{q}^T \mathbf{x}$$

  - Constraints:

$$0.488509x_1 + 0.063565x_2 + 0.945686x_3 \leq 2.865062,$$
$$-0.578592x_1 - 0.324014x_2 - 0.501754x_3 \leq -1.491608,$$
$$-0.719203x_1 + 0.099562x_2 + 0.445225x_3 \leq 0.519588,$$
$$-0.346896x_1 + 0.637939x_2 - 0.257623x_3 \leq 1.584087,$$
$$-0.202821x_1 + 0.647361x_2 + 0.920135x_3 \leq 2.198036,$$
$$-0.983091x_1 - 0.886420x_2 - 0.802444x_3 \leq -1.301853,$$
$$-0.305441x_1 - 0.180123x_2 - 0.515399x_3 \leq -0.738290,$$
$$x_1, x_2, x_3 \geq 0,$$

  where

$$\mathbf{Q} = \begin{pmatrix} 0.992934 & -0.640117 & 0.337286 \\ -0.640117 & -0.814622 & 0.960807 \\ 0.337286 & 0.960807 & 0.500874 \end{pmatrix}, \mathbf{q} = \begin{pmatrix} -0.992372 \\ -0.046466 \\ 0.891766 \end{pmatrix}.$$

- Global minimum: $f^* = -31.5285$
- Global minimizer: $\mathbf{x}^* = (5.210677, 5.027908, 0.000000)$

- **Test Problem Horst 7 [55]**

  - Number of variables: $n = 3$
  - Objective function:

$$f(\mathbf{x}) = -\left(x_1 + \frac{1}{2}x_3 - 2\right)^2 - |x_1 + \frac{1}{2}x_2 + \frac{2}{3}x_3|^{\frac{3}{2}}$$

  - Constraints:

$$-x_1 - x_2 + \frac{1}{2}x_3 \leq 1,$$
$$x_1 + 2x_2 \leq 6,$$
$$2x_1 + 4x_2 + 2x_3 \geq 1,$$
$$x_3 \leq 3,$$
$$x_1, x_2, x_3 \geq 0.$$

  - Global minimum: $f^* = -44.859$
  - Global minimizer: $\mathbf{x}^* = (6.0, 0.0, 2.0)$

References

1. Baker, C.A., Watson, L.T., Grossman, B., Mason, W.H., Haftka, R.T.: Parallel global aircraft configuration design space exploration. In: Tentner, A. (ed.) High Performance Computing Symposium 2000, pp. 54–66. Society for Computer Simulation International. San Diego, CA (2000)
2. Baravykaitė, M., Čiegis, R., Žilinskas, J.: Template realization of generalized branch and bound algorithm. Math. Model. Anal. **10**(3), 217–236 (2005)
3. Baritompa, W.: Customizing methods for global optimization — a geometric viewpoint. J. Global Optim. **3**(2), 193–212 (1993)
4. Bartholomew-Biggs, M.C., Parkhurst, S.C., Wilson, S.P.: Using DIRECT to solve an aircraft routing problem. Comput. Optim. Appl. **21**(3), 311–323 (2002). doi:10.1023/A:1013729320435
5. Belevičius, R., Ivanikovas, S., Šešok, D., Valentinavičius, S., Žilinskas, J.: Optimal placement of piles in real grillages: experimental comparison of optimization algorithms. Inform. Tech. Contr. **40**(2), 123–132 (2011)
6. Belevičius, R., Valentinavičius, S., Michnevič, E.: Multilevel optimization of grillages. J. Civil Eng. Manag. **8**(1), 98–103 (2002)
7. Björkman, M., Holmström, K.: Global optimization using the DIRECT algorithm in Matlab. Advanced Modeling and Optimization, **1**(2), 17–37 (1999)
8. Bomze, I.M., Eichfelder, G.: Copositivity detection by difference-of-convex decomposition and ω-subdivision. Math. Program. **138**(1–2), 365–400 (2013)
9. Breiman, L., Cutler, A.: A deterministic algorithm for global optimization. Math. Program. **58**(1–3), 179–199 (1993)
10. Bundfuss, S., Dür, M.: Algorithmic copositivity detection by simplicial partition. Lin. Algebra Appl. **428**(7), 1511–1523 (2008)
11. Butz, A.R.: Space filling curves and mathematical programming. Inform. Contr. **12**, 319–330 (1968)
12. Carter, R.G., Gablonsky, J.M., Patrick, A., Kelley, C.T., Eslinger, O.J.: Algorithms for noisy problems in gas transmission pipeline optimization. Optim. Eng. **2**(2), 139–157 (2001). doi:10.1023/A:1013123110266
13. Chandra, R., Menon, R., Dagum, L., Kohr, D., Maydan, D., McDonald, J.: Parallel Programming in OpenMP. Morgan Kaufmann, Los Altos (2000)
14. Chapman, B., Jost, G., Van Der Pas, R.: Using OpenMP: Portable Shared Memory Parallel Programming, vol. 10. MIT, Cambridge (2008)
15. Čiegis, R.: On global minimization in mathematical modelling of engineering applications. In: Törn, A., Žilinskas, J. (eds.) Models and Algorithms for Global Optimization. Springer Optimization and Its Applications, vol. 4, pp. 299–310. Springer, New York (2007)

R. Paulavičius and J. Žilinskas, *Simplicial Global Optimization*,
SpringerBriefs in Optimization, DOI 10.1007/978-1-4614-9093-7,
© Remigijus Paulavičius, Julius Žilinskas 2014

16. Čiegis, R., Henty, D., Kågström, B., Žilinskas, J. (eds.): Parallel Scientific Computing and Optimization. Springer Optimization and Its Applications, vol. 27. Springer, New York (2009)

17. Cox, S.E., Haftka, R.T., Baker, C.A., Grossman, B., Mason, W.H., Watson, L.T.: A comparison of global optimization methods for the design of a high-speed civil transport. J. Global Optim. **21**, 415–432 (2001). doi:10.1023/A:1012782825166

18. Csendes, T.: Generalized subinterval selection criteria for interval global optimization. Numer. Algorithms **37**(1–4), 93–100 (2004)

19. D'Apuzzo, M., Marino, M., Migdalas, A., Pardalos, P.M., Toraldo, G.: Parallel computing in global optimization. In: Kontoghiorghes, E.J. (ed.) Handbook of Parallel Computing and Statistics, pp. 225–258. Chapman & Hall, London (2006)

20. Delaunay, B.: Sur la sphere vide. Izv. Akad. Nauk SSSR, Otdelenie Matematicheskii i Estestvennyka Nauk **7**(793–800), 1–2 (1934)

21. Di Pillo, G., Grippo, L.: Exact penalty functions in constrained optimization. SIAM J. Contr. Optim. **27**(6), 1333–1360 (1989). doi:10.1137/0327068

22. Dickinson, P.J.: On the exhaustivity of simplicial partitioning. J. Global Optim. 1–15 (2012). doi: 10.1007/s10898-013-0040-7

23. Dixon, L., Szegö, C.: The global optimisation problem: An introduction. In: Dixon, L., Szegö, G. (eds.) Towards Global Optimization, vol. 2, pp. 1–15. North-Holland, Amsterdam (1978)

24. Dorsey, R.E., Mayer, W.J.: Genetic algorithms for estimation problems with multiple optima, nondifferentiability, and other irregular features. J. Bus. Econ. Stat. **13**(1), 53–66 (1995)

25. Dür, M., Stix, V.: Probabilistic subproblem selection in branch-and-bound algorithms. J. Comput. Appl. Math. **182**(1), 67–80 (2005)

26. Edelsbrunner, H., Grayson, D.R.: Edgewise subdivision of a simplex. Discrete Comput. Geom. **24**(4), 707–719 (2000)

27. Evtushenko, Y., Posypkin, M.: A deterministic approach to global box-constrained optimization. Optim. Lett. **7**(4), 819–829 (2013)

28. Ferreira, A., Pardalos, P.M. (eds.): Solving Combinatorial Optimization Problems in Parallel: Methods and Techniques. Lecture Notes in Computer Science, vol. 1054. Springer, New York (1996)

29. Finkel, D.E.: DIRECT Optimization Algorithm User Guide. Center for Research in Scientific Computation, North Carolina State University, vol. 2 (2003)

30. Finkel, D.E.: Global Optimization with the DIRECT Algorithm. Ph.D. thesis, North Carolina State University (2005)

31. Finkel, D.E., Kelley, C.T.: Additive scaling and the DIRECT algorithm. J. Global Optim. **36**, 597–608 (2006). doi:10.1007/s10898-006-9029-9

32. Fletcher, R.: Practical Methods of Optimization, vol. 37. Wiley, New York (1987)

33. Floudas, C.A.: Deterministic Global Optimization: Theory, Methods and Applications. Nonconvex Optimization and its Applications, vol. 37. Kluwer, Dordrecht (2000)

34. Floudas, C.A., Pardalos, P.M.: Encyclopedia of Optimization, vol. 1–6. Kluwer, Dordrecht (2001)

35. Gablonsky, J.M.: Modifications of the DIRECT Algorithm. Ph.D. thesis, North Carolina State University (2001)

36. Gablonsky, J.M., Kelley, C.T.: A locally-biased form of the DIRECT algorithm. J. Global Optim. **21**, 27–37 (2001). doi:10.1023/A:1017930332101

37. Galperin, E.A.: The cubic algorithm. J. Math. Anal. Appl. **112**(2), 635–640 (1985)

38. Galperin, E.A.: Precision, complexity, and computational schemes of the cubic algorithm. J. Optim. Theor. Appl. **57**, 223–238 (1988)

39. Gan, G., Ma, C., Wu, J.: Data Clustering: Theory, Algorithms, and Applications. SIAM, Philadelphia (2007)

40. Gendron, B., Crainic, T.G.: Parallel branch-and-bound algorithms: survey and synthesis. Oper. Res. **42**(6), 1042–1066 (1994)

41. Gergel, V.P.: A global optimization algorithm for multivariate function with Lipschitzian first derivatives. J. Global Optim. **10**(3), 257–281 (1997)

42. Goffe, W., Ferrier, G., Rogers, J.: Global optimization of statistical functions with simulated annealing. J. Econometrics **60**(1–2), 65–99 (1994). doi:10.1016/0304-4076(94)90038-8

43. Gonçalves, E.N., Palhares, R.M., Takahashi, R.H.C., Mesquita, R.C.: Algorithm 860: SimpleS – an extension of Freudenthal's simplex subdivision. ACM Trans. Math. Software **32**(4), 609–621 (2006)

44. Gorodetsky, S.: Paraboloid triangulation methods in solving multiextremal optimization problems with constraints for a class of functions with Lipschitz directional derivatives. Vestnik of Lobachevsky State University of Nizhni Novgorod **1**, 144–155 (2012)

45. Gourdin, E., Hansen, P., Jaumard, B.: Global optimization of multivariate Lipschitz functions: Survey and computational comparison. Les Cahiers du GERAD (1994)

46. Grbić, R., Nyarko, E.K., Scitovski, R.: A modification of the DIRECT method for Lipschitz global optimization for a symmetric function. J. Global Optim. 1–20 (2012). doi:10.1007/s10898-012-0020-3

47. Gropp, W., Lusk, E.L., Skjellum, A.: Using MPI-: Portable Parallel Programming with the Message Passing Interface, vol. 1. MIT, Cambridge (1999)

48. Hansen, E., Walster, G.W.: Global Optimization Using Interval Analysis, 2 edn. Dekker, New York (2004)

49. Hansen, P., Jaumard, B.: Lipshitz optimization. In: Horst, R., Pardalos, P.M. (eds.) Handbook of Global Optimization, vol. 1, pp. 407–493. Kluwer, Dordrecht (1995)

50. He, J., Verstak, A., Watson, L.T., Sosonkina, M.: Design and implementation of a massively parallel version of DIRECT. Comput. Optim. Appl. **40**(2), 217–245 (2008)

51. He, J., Watson, L.T., Ramakrishnan, N., Shaffer, C.A., Verstak, A., Jiang, J., Bae, K., Tranter, W.H.: Dynamic data structures for a DIRECT search algorithm. Comput. Optim. Appl. **23**, 5–25 (2002). doi:10.1023/A:1019992822938

52. Horst, R.: An algorithm for nonconvex programming problems. Math. Program. **10**(1), 312–321 (1976)

53. Horst, R.: A general class of branch-and-bound methods in global optimization with some new approaches for concave minimization. J. Optim. Theor. Appl. **51**, 271–291 (1986)

54. Horst, R.: On generalized bisection of n-simplices. Math. Comput. Am. Math. Soc. **66**(218), 691–698 (1997)

55. Horst, R., Pardalos, P.M., Thoai, N.V.: Introduction to Global Optimization. Nonconvex Optimization and Its Application. Kluwer, Dordrecht (1995)

56. Horst, R., Thoai, N.: Modification, implementation and comparison of three algorithms for globally solving linearly constrained concave minimization problems. Computing **42**(2–3), 271–289 (1989)

57. Horst, R., Thoai, N., De Vries, J.: On geometry and convergence of a class of simplicial covers. Optimization **25**(1), 53–64 (1992)

58. Horst, R., Tuy, H.: On the convergence of global methods in multiextremal optimization. J. Optim. Theor. Appl. **54**, 253–271 (1987)

59. Horst, R., Tuy, H.: Global Optimization: Deterministic Approaches. Springer, Berlin (1996)

60. Jansson, C., Knuppel, O.: A global minimization method: The multi-dimensional case. Tech. rep., TU Hamburg Harburg (1992)

61. Jaumard, B., Ribault, H., Herrmann, T.: An on-line cone intersection algorithm for global optimization of multivariate Lipschitz functions. Cahiers du GERAD **95**(7) (1995)

62. Jennrich, R.I., Sampson, P.F.: Application of stepwise regression to non-linear estimation. Technometrics **10**(1), 63–72 (1968). doi:10.1080/00401706.1968.10490535

63. Jones, D.R.: The DIRECT global optimization algorithm. In: Floudas, C.A., Pardalos, P.M. (eds.) The Encyclopedia of Optimization, pp. 431–440. Kluwer, Dordrecht (1999)

64. Jones, D.R., Perttunen, C.D., Stuckman, B.E.: Lipschitzian optimization without the Lipschitz constant. J. Optim. Theor. Appl. **79**(1), 157–181 (1993). doi:10.1007/BF00941892

65. Kearfott, B.: A proof of convergence and an error bound for the method of bisection in $\mathbb{R}^n$. Math. Comput. **32**(144), 1147–1153 (1978)

66. Kogan, J.: Introduction to Clustering Large and High-Dimensional Data. Cambridge University Press, London (2006)

67. Kolmogorov, A., Fomin, S.: Elements of Function Theory and Functional Analysis. Nauka, Moscow (1968)
68. Kreinovich, V., Csendes, T.: Theoretical justification of a heuristic subbox selection criterion for interval global optimization. Centr. Eur. J. Oper. Res. **9**(3), 255–265 (2001)
69. Křivý, I., Tvrdík, J., Krpec, R.: Stochastic algorithms in nonlinear regression. Comput. Stat. Data Anal. **33**(3), 277–290 (2000). doi:10.1016/S0167-9473(99)00059-6
70. Kvasov, D.E., Pizzuti, C., Sergeyev, Y.D.: Local tuning and partition strategies for diagonal go methods. Numerische Mathematik **94**(1), 93–106 (2003)
71. Kvasov, D.E., Sergeyev, Y.D.: Multidimensional global optimization algorithm based on adaptive diagonal curves. Comput. Math. Math. Phys. **43**(1), 40–56 (2003)
72. Kvasov, D.E., Sergeyev, Y.D.: A univariate global search working with a set of Lipschitz constants for the first derivative. Optim. Lett. **3**, 303–318 (2009). doi:10.1007/s11590-008-0110-9
73. Kvasov, D.E., Sergeyev, Y.D.: Lipschitz gradients for global optimization in a one-point-based partitioning scheme. J. Comput. Appl. Math. **236**(16), 4042–4054 (2012)
74. Kvasov, D.E., Sergeyev, Y.D.: Univariate geometric Lipschitz global optimization algorithms. Numer. Algebra Contr. Optim. **2**(1), 69–90 (2012). doi:10.3934/naco.2012.2.69
75. Křivý, I., Tvrdík, J., Krepec, R.: Stochastic algorithms in nonlinear regression. Comput. Stat. Data Anal. **33**(3), 277–290 (2000). doi:10.1016/S0167-9473(99)00059-6
76. Lanczos, C.: Applied Analysis, pp. 272–280. Prentice Hall, Englewood Cliffs (1956)
77. Lawler, E.L., Wood, D.E.: Branch-and-bound methods: A survey. Oper. Res. **14**(4), 699–719 (1966)
78. Leisch, F.: A toolbox for k-centroids cluster analysis. Comput. Stat. Data Anal. **51**(2), 526–544 (2006)
79. Lera, D., Sergeyev, Y.D.: Acceleration of univariate global optimization algorithms working with Lipschitz functions and Lipschitz first derivatives. SIAM J. Optim. **23**(1), 508–529 (2013)
80. Liuzzi, G., Lucidi, S., Piccialli, V.: A partition-based global optimization algorithm. J. Global Optim. **48**, 113–128 (2010). doi:10.1007/s10898-009-9515-y
81. Liuzzi, G., Lucidi, S., Piccialli, V.: A DIRECT-based approach for large-scale global optimization problems. Comput. Optim. Appl. **45**(2), 353–375 (2010)
82. Madsen, K., Žilinskas, J.: Testing branch-and-bound methods for global optimization. Tech. Rep. IMM-REP-2000-05, Technical University of Denmark (2000)
83. Madsen, K., Žilinskas, J.: Parallel branch-and bound attraction based methods for global optimization. In: Dzemyda, G., Šaltenis, V., Žilinskas, A. (eds.) Stochastic and Global Optimization. Nonconvex Optimization and its Applications, pp. 175–187. Kluwer, Dordrecht (2002)
84. Mayne, D.Q., Polak, E.: Outer approximation algorithm for nondifferentiable optimization problems. J. Optim. Theor. Appl. **42**(1), 19–30 (1984)
85. McMullen, P.: The maximum numbers of faces of a convex polytope. Mathematika **17**(02), 179–184 (1970)
86. Meewella, C.C., Mayne, D.Q.: An algorithm for global optimization of Lipschitz continuous functions. J. Optim. Theor. Appl. **57**(2), 307–323 (1988)
87. Meewella, C.C., Mayne, D.Q.: An efficient domain partitioning algorithms for global optimization of rational and Lipschitz continuous functions. J. Optim. Theor. Appl. **61**(2), 247–270 (1989)
88. Migdalas, A., Pardalos, P.M., Storøy, S.: Parallel Computing in Optimization. Applied Optimization, vol. 7. Kluwer, Dordrecht (1997)
89. Mladineo, R.H.: An algorithm for finding the global maximum of a multimodal, multivariate function. Math. Program. **34**(2), 188–200 (1986)
90. Mladineo, R.H.: Convergence rates of a global optimization algorithm. Math. Program. **54**(1–3), 223–232 (1992)
91. Mockus, J.: On the Pareto optimality in the context of Lipschitzian optimization. Informatica **22**(4), 521–536 (2011)

92. Moore, R.E.: Methods and Applications of Interval Analysis, vol. 2. SIAM, Philadelphia (1979)
93. Osborne, M.R.: Some aspects of nonlinear least squares calculations. In: Lootsma (ed.) Numerical Methods for Nonlinear Optimization, pp. 171–189. Academic, New York (1972)
94. Pardalos, P.M. (ed.): Parallel Processing of Discrete Problems. IMA Volumes in Mathematics and its Applications, vol. 106. Springer, New York (1999)
95. Pardalos, P.M., Hansen, P.: Data Mining and Mathematical Programming, vol. 45. American Mathematical Society, Providence (2008)
96. Paulavičius, R., Žilinskas, J.: Analysis of different norms and corresponding Lipschitz constants for global optimization. Tech. Econ. Dev. Econ. 12(4), 301–306 (2006)
97. Paulavičius, R., Žilinskas, J.: Analysis of different norms and corresponding Lipschitz constants for global optimization in multidimensional case. Inform. Tech. Contr. 36(4), 383–387 (2007)
98. Paulavičius, R., Žilinskas, J.: Improved Lipschitz bounds with the first norm for function values over multidimensional simplex. Math. Model. Anal. 13(4), 553–563 (2008)
99. Paulavičius, R., Žilinskas, J.: Global optimization using the branch-and-bound algorithm with a combination of Lipschitz bounds over simplices. Tech. Econ. Dev. Econ. 15(2), 310–325 (2009)
100. Paulavičius, R., Žilinskas, J.: Influence of Lipschitz bounds on the speed of global optimization. Tech. Econ. Dev. Econ. 18(1), 54–66 (2012). doi:10.3846/20294913.2012.661170
101. Paulavičius, R., Žilinskas, J.: Simplicial Lipschitz optimization without the Lipschitz constant. J. Global Optim. (2013, in press). doi:10.1007/s10898-013-0089-3
102. Paulavičius, R., Žilinskas, J., Grothey, A.: Investigation of selection strategies in branch and bound algorithm with simplicial partitions and combination of Lipschitz bounds. Optim. Lett. 4(2), 173–183 (2010). doi:10.1007/s11590-009-0156-3
103. Paulavičius, R., Žilinskas, J., Grothey, A.: Parallel branch and bound for global optimization with combination of Lipschitz bounds. Optim. Meth. Software 26(3), 487–498 (2011)
104. Pedoe, D.: Circles: A Mathematical View. Math. Assoc. Amer., Washington, DC (1995)
105. Petkovic, M.S., Petkovic, L.D.: Complex Interval Arithmetic and Its Applications, vol. 105. Wiley, New York (1999)
106. Pinter, J.: Extended univariate algorithms for n-dimensional global optimization. Computing 36(1), 91–103 (1986)
107. Pinter, J.: Globally convergent methods for n-dimensional multiextremal optimization. Optimization 17, 187–202 (1986)
108. Pinter, J.: Branch-and-bound algorithms for solving global optimization problems with Lipschitzian structure. Optimization 19(1), 101–110 (1988)
109. Pinter, J.: Continuous global optimization software: A brief review. Optika 52, 1–8 (1996)
110. Pintér, J.D.: Global Optimization in Action: Continuous and Lipschitz Optimization: Algorithms, Implementations and Applications. Nonconvex Optimization and Its Application. Springer, New York (1996)
111. Piyavskii, S.A.: An algorithm for finding the absolute minimum of a function. Theor. Optim. Solut. 2, 13–24 (1967). In Russian
112. Piyavskii, S.A.: An algorithm for finding the absolute extremum of a function. Zh. Vychisl. Mat. mat. Fiz 12(4), 888–896 (1972)
113. Sabo, K., Scitovski, R., Vazler, I.: One-dimensional center-based l_1-clustering method. Optim. Lett. 7, 5–22 (2013). doi:10.1007/s11590-011-0389-9
114. Scholz, D.: Deterministic Global Optimization: Geometric Branch-and-bound Methods and their Applications. Springer Optimization and Its Applications, vol. 63. Springer, New York (2012)
115. di Serafino, D., Liuzzi, G., Piccialli, V., Riccio, F., Toraldo, G.: A modified DIviding RECTangles algorithm for a problem in astrophysics. J. Optim. Theor. Appl. 151, 175–190 (2011). doi:10.1007/s10957-011-9856-9
116. Sergeyev, Y.D.: An information global optimization algorithm with local tunning. SIAM J. Optim. 5(4), 858–870 (1995)

117. Sergeyev, Y.D.: A one-dimensional deterministic global minimization algorithm. Comput. Math. Math. Phys. **35**(5), 553–562 (1995)
118. Sergeyev, Y.D.: A method using local tuning for minimizing functions with Lipschitz derivatives. In: Developments in Global Optimization, pp. 199–216. Springer, New York (1997)
119. Sergeyev, Y.D.: Global one-dimensional optimization using smooth auxiliary functions. Math. Program. **81**(1), 127–146 (1998)
120. Sergeyev, Y.D.: Multidimensional global optimization using the first derivatives. Comput. Math. Math. Phys. **39**(5), 711–720 (1999)
121. Sergeyev, Y.D.: Univariate global optimization with multiextremal non-differentiable constraints without penalty functions. Comput. Optim. Appl. **34**(2), 229–248 (2006)
122. Sergeyev, Y.D., Grishagin, V.: A parallel method for finding the global minimum of univariate functions. J. Optim. Theor. Appl. **80**(3), 513–536 (1994)
123. Sergeyev, Y.D., Kvasov, D.E.: Global search based on efficient diagonal partitions and a set of Lipschitz constants. SIAM J. Optim. **16**, 910–937 (2006). doi:10.1137/040621132
124. Sergeyev, Y.D., Kvasov, D.E.: Diagonal Global Optimization Methods. FizMatLit, Moscow (2008). In Russian
125. Sergeyev, Y.D., Kvasov, D.E.: Lipschitz global optimization and estimates of the Lipschitz constant. In: Chaoqun, M., Lean, Y., Dabin, Z., Zhongbao, Z. (eds.) Global Optimization: Theory, Methods and Applications, I, pp. 518–521. Global Link, Hong Kong (2009)
126. Sergeyev, Y.D., Kvasov, D.E.: Lipschitz global optimization. In: Cochran, J. (ed.) Wiley Encyclopedia of Operations Research and Management Science, vol. 4, pp. 2812–2828. Wiley, New York (2011)
127. Sergeyev, Y.D., Kvasov, D.E., Khalaf, F.M.: A one-dimensional local tuning algorithm for solving go problems with partially defined constraints. Optim. Lett. **1**(1), 85–99 (2007)
128. Sergeyev, Y.D., Pugliese, P., Famularo, D.: Index information algorithm with local tuning for solving multidimensional global optimization problems with multiextremal constraints. Math. Program. **96**(3), 489–512 (2003)
129. Sergeyev, Y.D., Strongin, R.G., Lera, D.: Introduction to global optimization exploiting space-filling curves. Springer, New York (2013)
130. Shubert, B.O.: A sequential method seeking the global maximum of a function. SIAM J. Numer. Anal. **9**, 379–388 (1972)
131. Snir, M.: MPI the Complete Reference: The MPI Core, vol. 1. MIT, Cambridge (1998)
132. Strongin, R.G.: Algorithms for multi-extremal mathematical programming problems employing the set of joint space-filling curves. J. Global Optim. **2**, 357–378 (1992)
133. Strongin, R.G., Sergeyev, Y.D.: Global multidimensional optimization on parallel computer. Parallel Comput. **18**(11), 1259–1273 (1992)
134. Strongin, R.G., Sergeyev, Y.D.: Global Optimization with Non-Convex Constraints: Sequential and Parallel Algorithms. KAP, Dordrecht (2000)
135. Todt, M.J.: The computation of Fixed Points and Applications. Lecture Notes in Economics and Mathematical Systems, vol. 24. Springer-Verlag, Berlin Heidelberg (1976)
136. Törn, A., Žilinskas, A.: Global Optimization. Lecture Notes in Computer Science, vol. 350. Springer, Berlin (1989)
137. Tuy, H.: Effect of the subdivision strategy on convergence and efficiency of some global optimization algorithms. J. Global Optim. **1**(1), 23–36 (1991)
138. Tuy, H., Horst, R.: Convergence and restart in branch-and-bound algorithms for global optimization. Application to concave minimization and dc optimization problems. Math. Program. **41**(1–3), 161–183 (1988)
139. Vaz, A.I.F., Vicente, L.: Pswarm: A hybrid solver for linearly constrained global derivative-free optimization. Optim. Meth. Software **24**(4–5), 669–685 (2009)
140. Watson, L.T., Baker, C.A.: A fully-distributed parallel global search algorithm. Eng. Comput. **18**(1/2), 155–169 (2001)
141. Wood, G.R.: Multidimensional bisection applied to global optimisation. Comp. Math. Appl. **21**(6–7), 161–172 (1991)

142. Wood, G.R.: The bisection method in higher dimensions. Math. Program. **55**, 319–337 (1992)
143. Yao, Y.: Dynamic tunneling algorithm for global optimization. IEEE Trans. Syst. Man Cybern. **19**(5), 1222–1230 (1989)
144. Zhang, B.P., Wood, G., Baritompa, W.: Multidimensional bisection: The performance and the context. J. Global Optim. **3**(3), 337–358 (1993)
145. Žilinskas, A.: On strong homogeneity of two global optimization algorithms based on statistical models of multimodal objective functions. Appl. Math. Comput. **218**(16), 8131–8136 (2012). doi:10.1016/j.amc.2011.07.051
146. Žilinskas, A., Žilinskas, J.: Global optimization based on a statistical model and simplicial partitioning. Comp. Math. Appl. **44**(7), 957–967 (2002). doi:10.1016/S0898-1221(02)00206-7
147. Žilinskas, A., Žilinskas, J.: Branch and bound algorithm for multidimensional scaling with city-block metric. J. Global Optim. **43**(2–3), 357–372 (2009)
148. Žilinskas, A., Žilinskas, J.: Interval arithmetic based optimization in nonlinear regression. Informatica **21**(1), 149–158 (2010)
149. Žilinskas, A., Žilinskas, J.: P-algorithm based on a simplicial statistical model of multimodal functions. TOP **18**, 396–412 (2010). doi:10.1007/s11750-010-0153-9
150. Žilinskas, A., Žilinskas, J.: A hybrid global optimization algorithm for non-linear least squares regression. J. Global Optim. **56**(2), 265–277 (2013). doi:10.1007/s10898-011-9840-9
151. Žilinskas, J.: Optimization of Lipschitzian functions by simplex-based branch and bound. Inform. Tech. Contr. **14**(1), 45–50 (2000)
152. Žilinskas, J.: Black box global optimization inspired by interval methods. Inform. Tech. Contr. **21**(4), 53–60 (2001)
153. Žilinskas, J.: Comparison of packages for interval arithmetic. Informatica **16**(1), 145–154 (2005)
154. Žilinskas, J.: Reducing of search space of multidimensional scaling problems with data exposing symmetries. Inform. Tech. Contr. **36**(4), 377–382 (2007)
155. Žilinskas, J.: Branch and bound with simplicial partitions for global optimization. Math. Model. Anal. **13**(1), 145–159 (2008). doi:10.3846/1392-6292.2008.13.145-159
156. Žilinskas, J.: Copositive programming by simplicial partition. Informatica **22**(4), 601–614 (2011)
157. Žilinskas, J., Bogle, I.D.L.: Balanced random interval arithmetic in market model estimation. Eur. J. Oper. Res. **175**(3), 1367–1378 (2006). doi:10.1016/j.ejor.2005.02.013
158. Žilinskas, J., Dür, M.: Depth-first simplicial partition for copositivity detection, with an application to MaxClique. Optim. Meth. Software **26**(3), 499–510 (2011). doi:10.1080/10556788.2010.544310